THE DOTTEREL

Desmond Nethersole-Thompson

THE DOTTEREL

COLLINS

St James's Place, London

William Collins Sons & Co Ltd
London · Glasgow · Sydney · Auckland
Toronto · Johannesburg

To Maimie, with love

First published 1973
© Desmond Nethersole-Thompson, 1973
ISBN 0 00 213057 2
Made and printed in Great Britain by
William Collins Sons & Co Ltd Glasgow

Contents

Photographs

Text Figures

Preface

ON a clear day stand beside the indicator on Ben MacDhui and look around. Near and far the great hills of Scotland are spread out before you. On just a few whalebacks, with gentle sparsely vegetated slopes or long stony ridges, dotterels live their summer lives. This grey wilderness is the dotterel's stage. Here I have watched these very special birds for over thirty years.

For the dotterel it is a woman's world. The hen is larger, brighter and more colourful. She courts and dominates, leaving her smaller duller mate to brood her eggs and rear their chicks. She also sometimes woos a second mate and gives him a clutch to brood. Afterwards she roams the hills with other 'grass widows'.

These are dynamic years for dotterels. They now sometimes nest on mountain plateaux in central and south-east Europe. In Holland these scarce and special mountain and arctic tundra plovers nest successfully on potato and beetroot fields in reclaimed polders below sea-level. Dotterels are again running on English Lakeland fells. This year, also, more pairs probably nested in Scotland than at any time for fifty years or more.

In Scotland the dotterel is not the only northern bird to prosper. A slightly cooler climate is one likely trigger. These last ten years or so, great northern diver and goldeneye, osprey, goshawk and snowy owl, green and wood sandpiper and Temminck's stint, fieldfare, redwing and bluethroat, have all nested, or tried to nest, on the Highland mainland or the northern isles. Whimbrels have already bred in Ross and Sutherland and black-tailed godwits in northern Caithness. Bonxies are spreading south. My own elusive snow bunting sings again in rough corries where the Victorians used to hunt them. Almost every year some new and exciting bird comes to nest in the Scottish Highlands.

This cooling climate also pleases skiers. Some now exploit our snowy corries for private profit. But surely we can conserve some whalebacks for our hill birds and those who love their wilderness.

Whatever happens in the future, these lovely and confiding dotterels will always call and hold us. We remember them with affection.

Desmond Nethersole-Thompson

Ross and Cromarty

CHAPTER I

The Dotterel

THERE is only one first time. It was 1st June, 1933 – a sunny morning with a strong southerly breeze – and a pair of dotterels were running over a great mossy lump of hill in the Grampians. As I lay watching them I could almost hear the warm voice of the old Irish naturalist saying that he envied me because I had not yet met dotterels on the hill.

On twinkling yellow legs they ran, stopped, and ran again over short grass and grey moss. My dotterels were quite close – little farther than the length of a cricket pitch – so I could see every feather. But the human senses are always selective, registering only a little. One dotterel, I saw, was slightly the larger and brighter, but that then meant little to me. What little gems they were with their short dark beaks and chocolate brown crowns. Broad white eye-stripes, circling their heads, met on the nape almost like a tennis player's sweat-band. When they dropped their heads they seemed to be wearing brown white-rimmed tam-o'-shanters. Cheeks and throat were white and upper breast dark-grey. Below this was another white band, almost like a ribbon of some illustrious imperial order. Then came the deep red or chestnut-brown lower breast and flanks and a great jet-black splash on the belly. Plump, compact, resplendent, and slightly smaller than lapwings, but how astonishingly cryptic they were! Each time one of them sat down or turned away, its ash-brown back and mantle melted into the grey moss carpet.

All the time the two dotterels twittered softly to one another, and before they flew away the larger bird lifted up its wings to show dove-grey underbanners. Then they were up, over the hill and far away, and all that was left with me were little whistling calls, fading away in the distance.

In the next fortnight I watched dotterels on the arctic wilderness of the Cairngorms. How I loved these hills, with their great rounded haunches and long stony flats and ridges. I did not then know it, but I had really had the luck of the devil. I had chosen one of the finest Junes in living memory – a year with plenty of dotterels on the hill. It was almost too easy! I was broken-hearted when I had to leave this wonderful country, but no doubt there would be other years. How little I knew what the dotterel and snow bunting would mean in the years ahead.

Back in the south I bought books and browsed in libraries and read everything about the dotterel that I could find. But how little I learnt. Even my friend, Jourdain, seemed puzzled. 'Nothing on record in the literature as to the courtship of this bird . . . Some evidence that one or two scrapes are made by one pair of birds, but we have no information as to which of the two is responsible for making the hollow . . . The only information as to the length of the incubation period is Heysham's oft-quoted assertion that it rarely lasts more than 18 or 20 days.'

We thus evidently knew little about the mating and courtship of this unusual bird, in which the hen was larger and brighter than the cock. Had this led to reversed courtship or possibly to polyandry? What territorial patterns had dotterels evolved? How did they choose their nest-sites? What were the respective shares of cock and hen in brooding eggs and rearing chicks? None of these questions appeared to have any real answers.

But the books certainly brought me down to earth. The Grampians and Cairngorms, to me so immense, were barely pinpoints on the dotterel's world range. My hills did contain the most regular westerly breeding groups in the world, but how pitifully small these were beside those concentrations on the fells and viddas of Norway, Sweden and Finland. In turn these groups were almost equally minute in the context of those mind-boggling tundras in the Soviet Union, where most of the world populations of dotterels bred. For the rest, there were only a few strays or pioneers in the Alps and mountains of Austria, Czechoslovakia and Romania.

But almost everything that the Old Naturalists wrote was exciting. In 1838 Thomas Heysham of Carlisle had written a famous paper on the haunts and manners of the dotterel which the Ornithological Establishment accepted as a classic. I recall the joy of reading it for the first time, but my heart was really in the dotterel's Highland heart-lands.

The dotterel had attracted some English eccentrics. In the 1860s and 1870s the Old Harrovian, Edward T. Booth, a dynamic alcoholic from Brighton, was at large among birds in the Highlands. A man of wealth, which he freely used to bribe and corrupt, Booth demanded the skins and eggs of every rare bird and dared God, Devil and Man to prevent him. 'You name 'em, I take 'em!' Erne and eagle, kite, harrier and osprey, dotterel, greenshank and diver! No detail omitted, and quite without humour, he described how all these birds had fallen dead at his feet! Ruthless, obsessional, and eccentric perhaps, but a naturalist extraordinary notwithstanding.

I also now read how H. W. Feilden and J. Harvie-Brown, two eminent Victorians, discovered a dotterel's nest on a spur of the

Grampians. 'She ran back behind the small grey stone where I saw her settle. Now I felt confident that I had her nest . . . She fluttered off her nest as if wounded and remained calling within 20 yards.' In my mind's eye, I could see Feilden marking the nest with his large red handkerchief and then excitedly trotting across the hill to Harvie-Brown and the gamekeeper who had guided them to triumph! 'She' and 'her'. How excited the two Victorians would have been had they known more about the dotterel's extraordinary pattern of reversed courtship and brooding!

The dream was there: I had only to realise it. By the end of March, 1934, I was back in the Highlands. In the winter I had taught in a school at Hertfordshire and skimped and saved enough money to give me spring and summer in the hills. It would be a shoe-string budget but I could hardly wait. I had already been on the move – after ravens in Wales and crossbills in Breckland. But this was quite different. Now, I thought, I was in the final. It was to be the jackpot or nothing. Eagles in April, greenshanks in May, and then dotterels and snow buntings all summer. Lean and strong, so keen and so confident, I had all the arrogance of my twenty-six years.

Carrie, who had first shown me dotterels in 1933, laughed at my fire-eating, but she already had a small green tent ready for the hill. In *The Snow Bunting* I told a little about that memorable summer and how the Cairngorms soon reduced the bold and bumptious young man down to size. But it was a wonderful year with everything to learn and so much that was new.

When we struck camp in August we had given pet names to some of our dotterels. We had watched Meeson and Sheila courting and mating and as partners in a kind of nest-dance. We had even watched Sheila laying her first egg after a strange ceremony in which Meeson had shared.

There was Ramsay, an endearing little cock dotterel, who was quite fearless, treating us like wandering red deer and sometimes flying off eggs and almost up into our faces. Ramsay, whose hen laid eggs with markings like Chinese hieroglyphics! Best of all, Blackie, who now nearly forty years on, is still my favourite dotterel. Blackie, the hen who ran with two cocks, leaving each with a clutch of her big black-blotched olive-brown eggs. Blackie, who sometimes relieved her second mate in the evening, and helped to brood the eggs. These, and some of the other dotterels in that wonderful year, are always remembered.

1934 was my honeymoon year with the dotterels. I only wish that I now understood as much about them as I then thought that I did! The first year of a new field project is particularly rewarding. Each week – almost every day – you learn fresh and unexpected facts. Everything seems so easy and fulfilling. But understanding and interpreting takes longer. Much longer. Each problem solved reveals more, usually more complex and equally intriguing. That is what research is about and that is why we do it. Eager young dotterel watchers will thus continue to discover fresh slants and new meanings. We have not yet reached the end of the beginning.

CHAPTER 2

The Acquisitive Society

DOWN the centuries the dotterel has been fighting a losing battle against insensitive man and engaging in an uneasy struggle with natural predators and competitors in harsh environments. For hundreds of years kings, nobles and commons have preyed on the dotterel. In 1510 a Steward of the Duchy of Northumberland wrote in the Household Book: 'Item dotterels. To be bought for my lord when they are in season to be had for jd apece.'

In the seventeenth century gangs of Norfolk wildfowlers hustled resting dotterels into nets. Lifting their arms, they 'chapped' stones together to frighten them into flight. Dotterels were also sometimes the sport of kings. At Thetford on 7th May, 1610, the slobbering King James I and his guests watched hares coursed on the heath and the Duke of Lennox flew a sparrow hawk at dotterels.

In the seventeenth and eighteenth centuries gourmets were more persistent persecutors than sportsmen but, on 16th August, 1786, Colonel Thomas Thornton, of Thornville Royal, in Yorkshire, took hawks and guns and proceeded up Glen Einich and then 'ascended' Sgoran Dubh in the Cairngorms. Between showers the tiercels killed a dotterel and three and a half brace of ptarmigan and the colonel shot the same number of both. 'Before my powder was quite wet I contrived to kill a dottrel; had my powder from the first been dry I could have killed 7 or 8 brace.'

Has any falconer before or since flown trained hawks at dotterels on the high tops?

But the dotterel was not merely a sporting bird in Strathspey. The Grant of Rothiemurchus, who 'kept the most enviable table in the world', often had dotterels as titbits.

In the late eighteenth century, about the time when the sporting colonel was exploring the Highlands, the fly fishermen of England were also after the dotterel. Their feathers were held in the highest estimation with anglers for making artificial flies. Thomas Bewick, of Newcastle, was brutally frank. 'The bird itself when stripped of its plumage sells only for 4d but its feathers at Keswick are always worth 6d.'

True naturalists were scarce in a trophy-hunting society which preferred dotterels in the hand to dotterels live and free, but there were a few exceptions. On 14th February, 1785, Dr John Heysham, J.P., of

Carlisle, was the first to describe a dotterel's egg in Britain. 'Some time last summer a nest of the dotterel was found on Skiddaw: the old one was killed and the eggs brought away, which were 3 or 4 in number. I saw three of them. They are somewhat larger than a magpie's egg; the ground is a dirty clay colour, marked with larger irregular black spots.'

The doctor's son, Thomas Coulthard Heysham, a shy and scholarly recluse, was a gifted all-round naturalist. 'A man of active habits, enjoying ample leisure, and inheriting a handsome competency', the younger Heysham seldom emerged from his niche in small town society, but had many correspondents, ranging from learned men-of-science down to thirsty and acquisitive fellmen and tackle-dealers with open palms and itchy trigger fingers. The dotterel always fascinated Heysham, who continuously sought skins and eggs by proxy. For years he exchanged letters with the bibulous and over-optimistic, occasionally procuring skins, but never those coveted eggs. At last, in 1834, 'The Carlisle Procurer' struck oil. On 27th May the Keswick mail coach transported two bulky hampers consigned to 'T. C. Heysham, Esqr, Carlisle'. A letter followed by the next post.

'Mr Heysham. Respectful Sir: According to my promise and agreeable to your request I forward five remarkable fine dotterel for you – one of which is alive at starting. They are very scarce this season, and, in consequence, have been very difficult to obtain. I got them yesterday, on a very High Mountain (called Great Gavel) . . . As I forward them today by the mail, I think there will be no fear of their taking any harm before their reaching you. Therefore I think it quite unnecessary for me to do anything with them.

'With regard to eggs, etc., unless some of these old Birds contain some I have not been able to procure any yet, but I have several Shepherds who are on the look-out for that purpose. Should they succeed I will lose no time in forwarding them to you.

'I will make no charge for the Dottrel – you may give me whatever you think them worth. I am Yours Respectfully, Wm. Camm, Keswick May 28th 1834.
P.S. Please to return the baskett by the Carrier as soon as convenient. W.C.'

I do not know how much of his 'handsome competency' Respectful Sir remitted, but on 9th June, 1835, Yours Respectfully still seemed anxious to please. 'Sir: I send you thees lines to let you know there is dotrels now if you have the noaton of coming up to Keswick – and I will go with you when it suits you.'

From H.Q. Carlisle, Heysham commanded Camm to visit all 'the smittle places'. But by 26th June the 'man in the field' was less hopeful.

'The eggs of the dottrel will be difficult to procure as there is a great scarcity of birds this year. But there shall be no want of exursion on our part to get them.' Respectful Sir had possibly forgotten to enclose 'soomat on account'.

At last Thomas Heysham stopped dithering and requested James Cooper to go to Helvellyn on his behalf. Starting early from Carlisle, Cooper tramped over thirty miles, reaching the top of Helvellyn late in the afternoon. Almost immediately he located a pair of dotterels, which he hunted until nightfall. Then he curled up and slept on the hill. Next morning, near the end of Squirrel Edge, the edge between Whiteside and Helvellyn, Cooper found the nest. Then, without further rest, he wearily walked back to Carlisle, where Heysham gloated over the eggs in the study.

A week later Cooper persuaded Heysham to accompany him to Robinson Fell. On 3rd July they saw dotterels but found no nests. But, on the 5th, Cooper flushed one from two fresh eggs. 'On quitting them it immediately spread out its wings and tail which trailed on the ground a short distance from us and then went away without uttering a single note.'

Heysham immediately 'boxed up' and 'prayed' Cooper to continue. Soon Cooper had found a young one. 'It rose up close to my feet and ran before me or I should never have seen it.'

Safely home, Maister Eeesham settled down to write a powerful piece. 'Anxious as I have been for several years past to secure the eggs of dotterel, it was not until the present year that I had the gratification of accomplishing an object I had for so long in view. After repeated excursions to the Lake District this summer for the express purpose, I was so fortunate as to obtain their eggs in two different localities.' In this classic on the 'manners' of the dotterel, the anonymous Cooper, 'a very able assistant', received only a gracious nod from the justice's son in the villa.

Over twenty years on, Cooper, now an established naturalist and curator of Warrington Museum, wrote in *The Zoologist*: 'Mr Heysham's account, as quoted, cannot be taken as a guide by those who intend to look for the eggs for nest there is none . . . Mr Heysham only saw the place where I had found the first egg, I believe, on record. This was on Whiteside.'

The Acquisitive Society was highly competitive. Heysham had been lucky to win the first prize. In May, 1830, William Hewitson, a distinguished ornithologist, had written: 'I was very anxious to meet with the nest and eggs of the dotterel which I was informed was to be found here. We only saw the bird as it was wheeling round us in the fog.'

Three years later Hewitson tried to pump Heysham. 'Do you know anything of its breeding in Cumberland? I saw one in Helvellyn at the end of May.' In the end Hewitson spent five consecutive holidays in Lakeland before he finally 'secured' a nest on Robinson.

In the 1840s and 1850s the early 'procurers' had many emulators – mostly shadowy figures like Bowe, Greenwell, and Greenup – whose names constantly appeared on data-labels.

Many now sought eggs and skins by purchase. In 1849 a party of bloods set out by carriage to deliver a clutch to John Wolley. As the conveyance rattled over the road, a well-dined enthusiast insisted on seeing the eggs. 'One egg was accidently broken by one of the party who put a finger in it at a jolt in the carriage as he was pointing to it.' The reception of the travellers *chez* Wolley is not recorded!

In the second half of the century the egg-greedy and trigger-happy continued to harry the dotterel. From 1860 onwards Francis Nicholson, a Manchester cotton merchant, frequently holidayed in Lakeland. A tall, craggy-faced man with a lantern jaw, Nicholson was a patron of Grasmere Sports, a judge of Cumberland wrestling, and a noted owner of fighting cocks. His beery scouts hunted for eggs in Cumberland and for skins on farms in west Lancashire. Nicholson got a few eggs in Lakeland, but I doubt whether this greatly harmed the dotterel. His skin-hunting agents were more sinister. 'In 1876 more birds were seen on arrival than usual but a number were shot. Jo Crosthwaite said he had shot 9 but Dixon Barnes only saw him with 5.' The same Dixon Barnes used a dog and string while hunting for eggs on the Keswick fells.

Nicholson had an almost paternal pride in the exploits of his lads in Lancashire. '2 May 1886: a man saw a lot of 5 ash dotterel, he called them, the common dotterel, near the new road at Chorlton. As it was Sunday he did not like to shoot them. He went at daybreak the following morning, and so did nine others, and shot 2 out of the 5. He wounded a third but, as the farmer was coming, he had not time to search for it.'

Between 1899 and 1903 Richards and Moorcroft, two Preston tackle-dealers, supplied him with twenty-one more dotterels, all shot on spring passage through Lancashire, at 2/9 to 4/- a head.

In England there were scarcely enough eggs and skins to go round. Even Booth made one unnecessary journey. In 1876, on his way to harry red kites in Strathspey, he broke his journey to hunt Crossfell in Cumberland. 'After a long search over all the likely ground near the summit we did not meet with a single bird, and it is possible they have now deserted these parts.'

The English trophy hunters were curiously late in discovering the Highlands; but in 1866 Booth was on to dotterels in Glen Lyon. On 5th June he saw nothing. 'Wind being too high.' On the 26th he returned to a hill which none of his rivals had suspected, and on which, to this day, he alone has found nesting dotterels. The entry in his diary is admirably terse. 'Sun hot. Self to look for dotterels. Got an old bird and three young but lost one of the old birds.' But in *Rough Notes* he wrote more dramatically. 'Having now reached a sufficient height to commence our search, the pony with the lunch hampers was made fast, and with the keeper and two ghillies we formed in line and made our way slowly over the ground in order to raise the birds.' A whistle from a ghillie. A dotterel was on the wing! Eggs or chicks were there for the taking. 'A downy nestling soon ran uphill.' Booth pounced on it like a cat. The second, then the third, chick moved. Booth quickly throttled them, barely pausing to poke them into his gamebag. 'Judging that the entire brood had now been taken the old bird was next easily procured.' The ogre now waited for the second dotterel to arrive. What an ungrateful brute it was. 'At last it was knocked down, but falling winged over a brae in a patch of rough ground where blue hares had scratched numberless holes in the turf, it succeeded in evading capture and no dog having been brought uphill, in consequence of the distance and heat of the weather, we were forced at length to relinquish our search.'

Booth was thus the first in the long queue of predatory Sassenachs who have since crossed the Highland line after dotterels.

Egging and skin-collecting was not restricted to Britain. In Lapland John Wolley's native collectors took at least two dozen clutches of eggs and often shot breeding birds off their nests. In the late 1850s an obscure Bohemian apothecary also liberated a bag of dotterels' eggs large enough to arouse the envy of the greediest oophile or even possibly to earn him temporary expulsion by the law-abiding elders of our present Jourdain Society.

Yet the eggs and skins that these colourful characters amassed were almost invisible items in the toll exacted by contemporary society. Wherever they lived or made landfall the wretched dotterels were slaughtered. In the 1840s Lancashire Lads 'shot hundreds'. Sporting tykes in east Yorkshire also had noses for brass. In the East Wolds notorious shooters boasted of killing fifty brace in a season. The wildfowlers and fishermen of Solway were equally quick on the draw.

On upland farms on the borders indigent, thirsty, or merely protein-hungry Scots wiped out entire spring trips. In the late 1850s James Purves, a border gamekeeper, sometimes shot ten to fifteen brace in a day. From behind stone dykes, colliers and farmers ambushed the dotterels, afterwards selling their bag for dainty dishes or dry flies. In south Perth ghillies and shepherds joined in the bonanza, often consigning a batch of dotterels and 'fifty couples of golden plovers' to The Breadalbane's mansion in London. The dotterels were 'especially appreciated and kept separate when sent away'.

On fells in Scandinavia, Lapps took hundreds in hare traps, and in

1884 gunners shot over ten thousand on the Jylland heaths. All over Europe dotterels made money. In November, 1873, hundreds were 'exposed for sale' at Valletta Market in Malta.

In the nineteenth century no one really knows how many thousands of dotterels were turned to coin in the hands of the bird-stuffers, poulterers, tackle-dealers and their agents. The poor mossfool was just one of many casualties in an age of ostentatious consumption. Without legal or moral impediment the Victorians mercilessly destroyed any animal that was edible, saleable, or even remotely competitive. Rich and poor then harried the dotterels. They shot them for the table or stripped their feathers for dry flies. They skinned their small corpses and mounted them in glass cases. They robbed their nests, blew their eggs and displayed them in cabinets.

By the end of the century only a handful continued to run over the English fells, but in northern Europe and the Scottish Highlands dotterels still survived.

CHAPTER 3

Eggers and Nesters

THE dotterel has always attracted nest-hunters. Beauty, rarity, tameness at the nest, attractive eggs, and wild and remote homelands all help to make it a beckoning and exciting bird. To me, the hills and fells of the dotterel country, with their strange names and fascinating traditions, were always a challenge. I was merely one of many who never rested until I had gone, seen, and been conquered.

On warm windless days, particularly on the mossy lumps and knobs in the central Grampians, dotterels' nests are sometimes easy to find. As you walk slowly over the soft moss, keeping your eyes fixed well ahead, you suddenly see a small object move about fifty to a hundred yards in front. Uncertain, you stop and raise your glasses. But long before you have time to focus them the object has melted into blind ground broken by innumerable little mossy tumps. Field glasses at the ready you continue to look ahead. For a brief moment a small head shows on the skyline and vanishes just as quickly. Ah, there it is again! This time you have it. A middling-sized plover is running fast. Head and neck are held up stiffly and its tail is also rigidly compressed. At last you see the broad white breast band and the rich red-brown breast and black belly patch below. No doubt now: it *is* a dotterel. Eagerly you glass it. Is it cock or hen? It looks a bit thin and rather faded. Its flank feathers are shaggy. Probably a cock. Surely there is a nest here.

It moves forward diagonally, backwards and forwards, then runs back again. It stops, starts, half-circles. Sometimes you lose it over the brow of the hill. You wait. You walk forward. You have it. It *is* edging nearer and nearer. Now you walk backwards; never taking your eye off it. Quite suddenly it takes fright and runs away. But this time you stay put. There it is back and jerking its head in a bibulous hiccough. It stamps or patters, pivots forward, snatches at some real or imaginary fly or spider. It must be close to the nest now. Suddenly it runs, almost races, over the moss. The feathers above the brood-patches puff out. It swivels, pokes downwards, then sits down, making little wiggling movements. Then, still wiggling a little, the dotterel settles down. How well its bright colours break up its outline! You barely see the white cheeks and the deep brown white-rimmed bonnet against that lump of granite. Choose your marks without lowering your glass. Then pick them up with the naked eye and go forward. Just ten yards before you reach him, the little beauty stands up and steps out. Then he shuffles away almost at your feet – his wings beating and his white-tipped tail trailing like a fan. And all the time the little dotterel squeaks and chitters like some small mammal in distress.

Is it really so easy to find a dotterel's nest? On a warm windless day with good light a dotterel is an easy mark when it is running over a comparatively stoneless carpet of moss and short grass. But first you have to find your bird. If dotterels are scarce you will probably walk many a mile and circle and quarter many acres without seeing or even hearing a single bird. In high wind or in thick mist or in sleet or heavy rain you always shiver and flounder, but seldom find a nest unless a dotterel flutters off at your feet. But rough weather and big country do not explain away all our failures.

Some cocks brood so closely that you continuously walk past their nests without flushing them. Others sometimes behave like golden plovers, skimming low over the hill long before you reach their nests. You then walk away without realising what has happened. Just to make your task a little harder, some of these brutes of birds stay away for over an hour and then skim back again. Only by chance do you then learn what you have missed. Another dotterel circles and runs, backwards and forwards, but always refuses to return and show you the eggs.

High numbers are also sometimes baffling. You believe, or allow yourself to believe, that the bird running over the ridge ahead belongs to the nest that you have already found. You must quarter the same big hill many many times before you can even roughly estimate its breeding numbers. In June, 1949, for example, we found nine nests with eggs and two broods on tops and ridges which we knew well from previous years. We found three nests on the 9th, three more on the 12th and 13th, and three nests with eggs and a brood of small chicks on the 20th. Finally, on 11th July, we found a cock and another brood. Yet on other days we thought that we had thoroughly searched all the likely ground. Only the most naïve ever believe that they have found every nest in any part of the Cairngorms.

What know they of dotterels who only Drumochter know?

But we have had our days. Days when dotterels have run off their nests fifty to a hundred yards ahead and then ran back to them within a quarter of an hour! On 7th June, 1953, in less than six hours, two friends and I found four nests. Yet we should have done better. That day we missed at least five other nests.

On 5th June, 1941, in the morning, I watched Carrie find a greenshank's nest in the Cairngorm foothills. And in the afternoon she found three dotterels' nests on the high tops. I myself found nothing. Again, on 9th June, 1949, we had a day on the Cairngorms which I still find almost incredible. Between one o'clock and ten past eight we watched three dotterels and a hen snow bunting go back to nests and saw a second hen snow bunting building. Yet one of the dotterel's nests was a real 'teaser'. The cock ran and flew for nearly an hour before going 'down'. Just seven minutes after we had found that nest we saw another dotterel running. But this bird was quite different. He only took eighteen minutes to return to his eggs. A day like that happens only once in a lifetime.

For Brock, 4th June, 1961, was another 'dotterel day' in the Cairngorms. Cloud and rain delayed his start until 10 o'clock, but he made a good climb up to the dotterel country. About 12.30 he saw a cock running over the skyline in a stony shoulder, but lost it. Half an hour later he watched it run back and then fly about fifty yards to a nest with three beautiful eggs. Two hours later, on another ridge, he saw a cock join two hens and then all three dotterels flew away. Three minutes passed and then the cock skimmed low over the hill. In a couple of minutes he ran to a nest between two big stones. Mist now rolled up. First it rained, then there were bitter blinding showers and scurries of sleet. Brock almost gave up. But, as sometimes happens, the wind changed and the mist lifted. Brock now moved on and started to hunt a big flat. Suddenly, about a hundred yards ahead, he saw a cock running along with spread tail and flapping wings. Time after time it ran, jerked its head, chittered and 'distracted', but never would it show him the nest. Then the clouds dropped and everything was cold, wet and grey. Brock hung on, often losing the dotterel in the mist. At last, it edged slowly towards him and for a moment sank down beside a grey stone. Then it rose and flew away. Brock walked to the stone and beside it were three eggs.

I also recall other days of less success, but of greater fulfilment. From 6.30 to 10.15 on 3rd June, 1942, Carrie and I battled with a suspicious dotterel which just refused to go back. We had already missed another dotterel in bad light. Now we were really up against it. We tried everything – watching, searching, going away and returning – but the little dotterel would not play. He kept on running in circles, occasionally calling softly and jerking back his head. Yet it was a still warm evening. We had to watch for nearly four hours before he suddenly ran about a hundred and fifty yards and plumped down on a nest and then immediately started away again. In bad weather we should have had no chance.

A few of the legendary egg collectors have written about their dotterel hunting, but Booth was the only Victorian who made light of it.

'The nest is by no means difficult to find. . . . In making this assertion

I am aware that my experience differs from that of the majority of
observers. I have only on two occasions sought for nests containing eggs,
and both, after waiting and watching for an hour or two, were discovered. Patience and a slight knowledge of the habits of the birds
being all that is needed to ensure success.'

His MS Diaries, however, contain no account of his finds.

On 16th June, 1873, H. W. Feilden and J. A. Harvie-Brown are
usually credited with the 'first authenticated and carefully identified
Scottish nest and eggs'. I can picture the ending to a perfect day.
Carrying his box hat, with the precious eggs well-padded in moss,
Feilden walks gingerly downhill. Between finger and thumb, like a boy
with mushrooms, Harvie-Brown holds the red handkerchief containing
the nest-scoop and, walking respectfully, just two paces to the rear, the
Highland gamekeeper thinks hopefully of what he is about to receive.
Then, as the two heroes board the toy train, and puff off to Perth in the
mail van, the keeper, enthusiastically waving his cap, acidly assesses the
pedigree of the departing 'shentlemen' by what they have left in his
other hand.

In the late 1890s Macomb Chance evolved new skills and designed a
sheepskin coat to baffle the dotterels. On 15th June, 1898, he and two
Drumochter gamekeepers found three nests. At the first 'I donned my
special sheepskin coat and took up a position about 50 yards from
where she ran. In half an hour she appeared on the skyline and
gradually worked up . . . Although I was in full sight she did not seem
to notice me. No doubt this was due to my coat.'

His younger brother, Edgar, now chiefly remembered for his
research on the cuckoo, was another great nester who also loved
hunting for dotterels. Some of his hunts were extremely successful.
On 28th May, 1929, Chance and Douglas Meares, in four exciting
hours, found four dotterels' nests, all within a radius of a quarter of a
mile. The first nest they found at 12.5 p.m., the second at 1.50, the
third at 3 o'clock and the fourth and last at 3.55 p.m. 1929 was a good
year for dotterels in the central Grampians, but this was a fine feat of
nest-finding by any standards.

What new facts did the Chances discover about the dotterel?
Macomb was possibly the first to realise that the two dotterels were
seldom seen together close to a nest with eggs. 'I never found a nest if
I found both birds.' Edgar recognised that the colours and plumage
of cocks and hens were subtly different, but to the end he confused them.
'The ring of white below the breast is not so broad as in the case of the
cock, and the hen's back is more mottled. His appearing plainer.'

In 1922 Norman Gilroy vividly described the wilderness of the
Drumochter hills and how seven different dotterels behaved at their

nests. At the first 'the little creature was behaving in the most piteous fashion, running backwards and forwards all round me and sometimes actually between my feet, her wings, drenched with the mist, trailing on the ground and her beautiful white-tipped tail spread out like a fan. All the time she cried distressingly a plaintive querulous squeak of utter abandonment and misery.'

Since 1900 the north country has produced some dedicated enthusiasts. Between 1905 and 1912 J. F. Peters, a Lancashire cotton merchant, found six nests and three broods. Here are the laconic entries in his 1911 diary.

'28 May. Buttermere Fells. 2 hard-set eggs.

'4 June. Buttermere Fells. 2 hard-set eggs. This nest much neater and better lined than one of 28 May. The nests were rather over 100 yards apart. The bird on the second nest was much brighter and better marked and would not come nearer than 3–4 yards of us. Only saw single bird at each nest. On 21 May the birds of the second nest were together (though the nest was not found) and after running about for some time they flew away to the next mountain.

'11 June. Whiteside. Helvellyn. 3 eggs near hatched. The chicks chirped loudly when removed from the shell 27–28 hours after being taken.' No damned sentiment about Mr Peters.

In 1925 young Ernest Blezard, of Carlisle, was searching the fells for the third year. On 6th June he slept out in an old sheepfold. Next morning he spun a halfpenny – all that he had – to decide which fell to search. What a lucky spin that was! 'I was on the way back to the tops as soon as the rising sun began to expel the morning mists.' At 2,500 feet a dotterel darted across his path. Within the hour he was sitting beside the nest, recording the dotterel's behaviour.

The Old Guard – F. C. Selous, the lion hunter, the Chances, Gilroy and others in Scotland, and Peters, Baldwin Young and Bolam in England – are gone, but there are younger and equally dedicated watchers.

To-day Derek Ratcliffe has watched dotterels in many British habitats. On 29th May, 1959, he found his first English dotterel's nest on a traditional fell. 'I returned to the place where I had first seen the dotterel, intending to search the ground, but soon noticed the bird was following me and was running about 300–400 yards away. I again lay down to watch. It ran for a short distance, then doubled back, it hesitated over a stopping place, looked down and then settled, shuffling as it did so. I rose and made for the place, only about 30 yards away. The dotterel jumped at about 10 yards and promptly did the wing-trailing act. I was delighted to see two eggs lying on one of the hummocks on the highest part of the summit plateau.'

In 1967 Donald Watson found one of the few dotterels' nests ever recorded in the Southern Uplands. 'I worked slowly from the summit cairn and had almost decided to go no farther when I saw the dotterel running and feeding below me. It vanished; and I waited but nothing happened. So, after a brief look round, I went forward and searched the ground. Quite soon the dotterel shuffled off almost at my feet. There were the eggs, greenish-grey with dark splodges, in a neat round depression in fairly short grass.'

Sandy Tewnion is another outstanding montane ornithologist. So far he has found 32 nests and has taken fine photographs in the Cairngorms and Grampians.

And there were others. In 1953 a famous nest-finder 'walked-up' some likely dotterel ground. 'Suddenly I saw a dotterel running ahead, clearly off eggs. I kept my eyes glued to the bird, and took off my pack to mark the spot where I had first seen it. Then I walked away and watched it run back to my pack. Slowly he circled it. What fools these humans are! There was my pack, right on top of three beautiful but squashed eggs! There's marking for you!'

Egg-hunting dogs also belong to the dotterel saga. The nose of his labrador led Steward of Windermere to more than one nest on Helvellyn. One of Adam Watson's famous Nature Conservancy pointers also once – just once! – 'pointed' a dotterel's nest.

Have these colourful eccentrics really made the mossfool scarcer in Britain?

In the nineteenth century the egger and his touts were almost insignificant factors in the sinister complex of persecution. In an age when men boasted of their possessions, I have records of fewer than fifty clutches taken in England and Scotland. Surely not a deadly rate of predation.

In this century trophy-hunters and gunners have seldom operated, but eggers have continued, at first openly, later underground.

What are *the facts*? Between 1900 and 1970 egg-collectors have possibly taken 300 to 350 clutches in Scotland and England. I actually have records of 209 clutches from Scotland (many now in public or national collections) and, between 1900 and 1945, twenty-four clutches from England.

In a context of an average breeding population of about seventy pairs has this predation had lasting local or national significance? In the first quarter of this century the rape of two dozen clutches was possibly a factor in the final collapse of the fringe breeding groups in England. But more fundamental causes led to the breakdown.

In Scotland egg collecting appears to have had even less impact. In the 1950s and 1960s the average number of pairs nesting on the better-known haunts in the Cairngorms and Grampians has probably exceeded the mean in the 1920s and 1930s.

In 1967 Derek Ratcliffe wrote: 'In the case of rare species such as the hobby and dotterel, it is clear that, despite lurid predictions for many years, the taking of hundreds of clutches of both birds during the last fifty years has had no detectable effect on the size of the British breeding population.'

CHAPTER 4

Arrival, Courtship and Display

Arrival

THE earliest dotterels usually return to the Cairngorms in the first week of May – some of the most exciting days in the Highland year. In the valley greenshanks are laying in bogs and clearings and sandpipers 'wickering' beside lochs. In the old pine forests crested tits and siskins are brooding and at dawn blackcock and capercaillie leks are full of sound and colour. But the high tops to which the dotterels are homing are usually still deep in winter snow and large rocks and boulders are often the only black objects in a white world. How desolate and lifeless these great tops now are. Ptarmigan and an occasional eagle are almost the only birds on the move.

I have never seen dotterels on the low ground. They probably reach their nest tops by flying over the hills at night but, in early spring, gales, snow and mist always hamper the observer. One day you toil up to the tops and find no dotterels. A day or two later one, two, or a small trip are there. My first sightings for the Cairngorms range from 2nd to 12th May, but these refer to particular hills and are not valid for the whole area. There are at least two earlier records. On 12th March, 1961, W. Sinclair photographed a solitary dotterel in the central Cairngorms, and on 28th April, 1955, Adam Watson and Bruce Forman saw a pair on Glas Maol at 3,300 feet.

A few hens sometimes return ahead of small trips. In six years the earliest dotterels I saw were mixed groups. Small parties of hens were first on the hill in two years, and pairs, or solitary hens, in the remaining four years. In Britain dotterels apparently seldom travel in pairs.

J. Watson (1888) suggested that 'in Westmorland the dotterel makes its appearance on the fells on 8–15 May'. There are few recent English records, but on 12th May, 1948, Irvine Smith watched eight dotterels joisting on a hill in Lakeland. The unpublished diaries of Francis Nicholson, written in the late nineteenth century, also describe the unfortunate spring trips which formerly visited farms in Lancashire. On 17th April, 1890, for example, Nicholson received two dotterels, recently shot. 'They had very little sign of their summer plumage. This is a very early date. First week of May usual date.' On 25th April, 1893, Richards, a Preston gun-maker, 'dispatched' five dotterels shot near Garstang. 'One is in summer plumage and four have no sign of summer plumage.' On 2nd May, 1886, Nicholson received two hens 'in fine plumage'. On 3rd May, 1888, there were three hens 'in fine feather' and two cocks 'not yet fully assumed their summer dress', but on 9th May two cocks were 'in immature plumage'. How accurately did the 'procurers' sex these birds?

In Fenno-Scandia dotterels tend to arrive later. On Värriötunturi Fell (67° 45'N) E. Pulliainen (1971) found that they arrived before the snow had disappeared at the end of May – 30th May, 1969, and 25th May, 1970 – and began to court immediately. Within a week the earliest hen had laid an egg.

Pairing and Courtship

Climate and weather, particularly snowcover on the tops, largely control the tempo of courtship. Dotterels normally pair from groups. Exceptionally, however, a solitary hen establishes a rough display centre. On 12th May, 1946, for example, we watched a solitary hen displaying from a ridge close to a snowfield. Time after time she called and made advertising flights. Winnowing and gliding over the tops, she kept returning to the same ridge, but no other dotterel answered. This early display by a solitary hen is unusual and probably due to absence of others. I have only twice seen this happen, but lone or immature hens possibly prospect and squat in marginal habitats.

Dotterels usually mate on flats and plateaux. These small groups are often extremely tame. Hens raise their wings as they chase and try to isolate the cocks. A hen also runs ahead and sits down as if brooding and then runs on again. In the Cairngorms and Grampians, mating groups have consisted of five to eight birds. For example, on 12th May, 1942, I watched a group of four hens and three cocks on a Cairngorm top. Time after time, the hens raised their wings and shut them again. Then one suddenly ran forward, squatting and apparently brooding. When no cock followed her she returned to the group, raised her wings and soon afterwards again ran and squatted. Finally, all seven dotterels, whirring noisily, flew away and resumed display on a flat about three hundred yards ahead. In the Grampians, on 25th May, 1967, Colin Murdoch watched another group of seven, of three or four hens, and probably three cocks. A cock and hen left the group, the hen started to scrape and then ran a few yards forward. In the group two hens raised their wings and ran at one another like blackcocks at lek. The hens usually raised and held up their wings, but one cock also apparently did so.

If the sun is shining, and the sky is clear, the bleak and stony hillside with its great snowfields becomes a colourful tourney ground. In time the pairing groups break up as the new pairs hive off. Sometimes – particularly in years with more hens than cocks – two or more hens compete for one cock. Or, perhaps, an extra cock pursues a newly-formed pair. The three dotterels, 'skirring' excitedly, then fly high and low over tops and ridges, but continue to fight, call and display whenever they land.

Dotterels sometimes have different displays. On 21st May, 1945, for example, Perry watched four of them, hump-feathered, reeling and trilling like angry turnstones. Two cocks jumped over one another, 'one getting a feather in passing and then jumping on to the rump of a big hen, who bowed forward while the cock mated with her.' Between these spasms, uttering soft melodious calls and occasionally thin linnet-calls, the hens ran at high speed, dipping forward just before they met in physical contact. Trilling loudly, two hens humped themselves bill to bill, one charging and chasing away an attendant cock. The cocks also fought.

On 23rd May, 1970, Tewnion also watched six dotterels sparring and displaying. While one pair was nest dancing a dunlin persistently approached the hen dotterel. A second nest dancing pair fought with another pair. 'The cock of one pair fought the hen of the second and vice-versa.'

Late and early summers

Weather and climate rather than the calendar determine the dotterel's breeding-rhythm. In 1934 summer came late to the Cairngorms. On 27th May we watched a courting trio and two trips which had not dispersed. The dotterels flew swiftly over the snowy wilderness, seldom staying long on any ridge. On 7th June they were paired, but still without firm territories. In the afternoon I saw four small specks above the cliffs of a large rounded hill. Suddenly they flashed down, swerving, veering and fighting. Then they swished across the stones and lichen. There were four birds – two cocks and two hens. Almost immediately, they dashed at one another, all fury and passion. Feathers indrawn, backwards and forwards they ran, cock against cock and hen against hen. Sometimes they half spread their tails, showing white tail-tips. Or they met and clashed and, raising their wings, leap-frogged over one another. Then they turned and started again. All this action was fast. Within a minute one hen rose and flew over the hill shoulder, giving the rhythmic *peep-peep-peep* and winnow-glide of ceremonial flight. Then, a cock was away, flying so quickly that I could hardly follow him. The second pair – each bird in turn and independently – then made scrape movements, but without leaving any marks. Next morning the four dotterels were fighting again, but one pair finally dominated the other.

In other years dotterels have been equally late or later. After the severe winter of 1951, for instance, deep snow covered the high ridges until late June. On 15th June, with only a few snow-free patches among the grass-moss communities, Adam Watson saw six dotterels on the top of the highest hill. On the 19th, however, they were in pairs. 'The display that I saw consisted mostly of aerial chasing and some ground chasing. Once a hen ran rapidly after a cock. At the end of the chase she gracefully raised her wings and fanned her tail, but did not bow for copulation.' On 22nd June, six or seven pairs were displaying on snowy and on snow-free ground.

On 25th May, 1959, Brock watched two mating groups on the Cairngorms and on 27th May, 1939 – a late summer on the Drumochter Hills – I watched a hen – in a group of five – hold her lovely wings above her back before chasing and trying to separate a rather pathetic little cock, which ran quietly over the snow and then flew off in twisting flight, the hen behind him.

In years with exceptionally heavy snowfalls in May, trips and pairing groups are sometimes forced back into flocks. In the third week of May, 1954, for example, much snow fell on the Cairngorms. Early on 22nd May Watson saw 'a flock of nearly 40 dotterels flying northwards at close range, immediately below heavy banks of cloud and through veils of light-falling snow'. The hills were then snowbound down to 2,300 feet. In May, 1955, there were also heavy snowfalls. On 14th to 15th May the snowline was continuous to 2,000–2,500 feet and light cover to 1,400 feet. On 22nd to 23rd May snow was also down to 1,800–1,900 feet, the MacDhui–Cairngorm plateaux being completely covered on 23rd, except for a few boulders on the ridges and on the 30th 90% of the plateaux was almost like an ice-cap. On a walk from MacDhui to Cairngorm and back by a different route, the Watsons saw no dotterels. With heavy and continuous snowfalls in the Cairngorms from 11th May until the first thaw on the 23rd, the dotterels were presumably all kept off their breeding ground for about a fortnight. 1955 was certainly an exceptionally 'late' May, with a most unusual concentration of winter snow, but there have also been other years – like 1953 and 1968 – in which the dotterels were denied access to their normal habitats on the highest Cairngorm plateaux. In some of these years the hard weather probably forced some dotterels from the Cairngorm groups to nest farther south – possibly to the west Grampians.

In warm summers the pattern is different. A cock and hen dotterel paired, but still courting, and with no fixed territory, are restless and almost continuously on the move. In mild Mays dotterels sometimes

court, pair, join in nest dances, and lay their first eggs, all within a
week. For instance, on 11th May, 1961, Brock came upon an animated
group of three hens feeding together on a squashy patch of moss and
grass, not far from a large snowfield. Full of life and excitement, they
tinkled – those pure little whistles which carry so far in the windy
desolation of ben and tundra. Other hens called noisily beside a burn
and another rose from a ridge and flew in advertising-flight across the
Lairig Ghru, thus moving from massif to massif. Yet between 22nd
and 24th May at least three cocks had mated and already were down on
eggs.

Sexual behaviour after pairing

In birds with a reversed courtship, the adjustments and synchronising
of male and female sex-rhythms is a subtle process. At first the hen,
whose sex drive is initially stronger, dictates the pattern of early
ground-chases and displays, simultaneously isolating, dominating and
courting the cock, and attacking other hens. She also often chases the
cock or hen and then flies after it. Then, the two dotterels, flying fast,
sometimes become partners in a flight-dance, turning in harmony or
crossing tracks as they fly. Or, slowly and gracefully they flap their
wings like cock ringed-plovers in their display-flights. The cock also
sometimes follows the hen, flying almost as deliberately as an owl, his
legs sometimes dropped and dangling. At first cock and hen often
fight. Unmated cocks and hens also strive to possess those of others.
You are watching two birds, then you suddenly see three flashing over
ridge or flat and almost always you hear angry strident trilling. Cocks
soon become active partners in the sex-dances, in which they now
pursue and dominate the hens. Two cocks also now sometimes compete
to possess the same hen.

Between pairing and laying, scrape ceremonies or nest-dances help to
hold the pairs and probably synchronise their sex-rhythms. At first
either cock or hen makes scraping or nest-making movements independently; later both dotterels work in the same place. Either dotterel
starts the scraping movements and then calls up its mate. In pauses
between nest-dances, the dotterels often mate quite close to their
scrapes.

Let me now describe the mating of Meeson and Sheila. On 7th June
they were rotating and scraping, shuffling from tussock to tussock, first
one, then the other, leading. They were then only making the movements: they left no marks or hollows on the ground. In the next three
days Meeson and Sheila were finally settling on a territory where they
continued their nest-dances. At noon on 11th June, however, round and

fluffy like a child's woollen ball, Sheila was as broody as a farmyard hen. Time after time she sat down between two stones. So quiet and furtive were the two dotterels – even on that bright June day – it was difficult to find them. Up and down the stone-shot hill they ran; Sheila sometimes so broody that we either lost her or thought she was sitting. Whenever we moved she rose and ran away. Every now and then one of the dotterels *tinged* sharply to keep in touch with the other. Twice, Sheila sat between stones, twitching her tail and seemingly brooding. Each time Meeson ran almost up to her, uttering a jumble of sweet linnet-like notes. Carrying himself low, he sometimes stopped, made pecking movements, and seemed afraid to advance. At last, he ran up to her and gently billed or nibbled her nape. Then, as she sat between the stones with her tail slightly raised, he quickly possessed her, stepping on to her back and lowering his body on to hers, without even raising his wings. Five minutes later he had her again, but just before sinking on to her back, Meeson raised both lovely wings and held them so during copulation.

After Meeson had taken her for the second time, Sheila ran to a tussock in which, as we later found, there was a deep bowl. Thus she showed him the nest of her choice.

Dotterels have many different ways of mating. On 11th June, 1940, we watched a pair courting in the Cairngorms. *Tinging* sharply, the cock ran quickly over the stony flat, giving the soft rippling *tee-hee-hee* courtship cries. Then, breast to ground, he suddenly started to shuffle in half-circles, his feet jerking spasmodically. Standing in the scrape, he built sideways tucking in straw-coloured grasses. Then he slowly walked away from the nest, occasionally throwing scraps of moss over his shoulders towards it. Then he called up his hen, which now took a turn at shuffling and scraping. This ploy they repeated several times and then, almost as if some internal clock had run down, the two abandoned the scrape and ran forward. Again and again they paused then started nest-dancing, until the hen suddenly sat down beside a small green stone. There – without fuss or ceremony and without raising a wing – the cock almost casually served her. Passion now spent, the pair ran over the flat, the hen puffing out her breast feathers and preening herself. Then, suddenly lifting her wings, she stopped and almost instantly flew away. All this happened under a hundred yards from a nest on which another cock was sitting on three heavily-blotched, but deeply incubated eggs. Ten days later we found a nest with three fresh eggs close to where the pair had mated. Both these clutches had eggs of unusual shape and marking, suggesting that the same hen had laid them. We had thus possibly seen the hen accepting a second mate.

On 10th June, 1938, a rough and misty day, I worked the tops for over sixteen hours. Late in the evening, after the mist had lifted, I sat down on a broad slope overlooking a stony flat. While I was resting a pair of dotterels flew in and landed quite near me. Tilting forward at an angle of about 45° the hen ran over the flat just ahead of the cock, showing off her creamy-white and cinnamon undertail feathers. About fifty yards she ran like this, the cock calling softly a few yards behind her. Slower and slower she ran until she suddenly stopped and then gradually upended, breast almost to ground and body and tail cocked at an angle of about 60°. Behind, moving in stops and starts, the cock ran up and almost casually stepped on to her back. For a split second during the mating, the hen twisted her head and the two beaks touched in a 'dotterel's kiss'; then he clumsily shuffled from her back to the ground. The hen preened, fluffed herself, and sat in a grassy tump as if she was brooding or laying an egg. For less than a minute she stayed there, then she shook herself and ran a few yards, looking sleepy. Then she opened her wings and flew away, the cock slowly flapping behind her. The two dotterels suddenly increased speed and began to cross tracks. Flying faster and faster, and tinkling softly, they now disappeared over the hill.

These two components of the hen's pre-copulatory display – squatting and upending – are also shown by subordinate or pursued birds in sexual or territorial fighting. These actions thus presumably assist the cock to pursue and dominate.

Cock dotterels are sometimes lusty. On 2nd June, 1950, Carrie watched a cock serve his hen three times in just under five minutes. Balancing himself, wings high above his back, he did not wave or beat them during copulation; but at least once he raised his wings and fanned his tail just before he stepped on to her.

Courting a mate, a hen sometimes uses dove-grey underwings to excite him. On 24th May, 1952, a pair swished in and landed within a few yards of us. The hen now squatted down between two stones and repeatedly flashed her silvery underwings, half-raising and then abruptly shutting them. The cock, meanwhile, stood just behind, but did not mount her. These two lovely birds then flew off.

On 8th June, 1961, Brock watched a hen lying on a grass mat with her wings spread out. The cock ran up to her but, instead of mounting, he, too, crouched flat almost beside her. A few seconds later the two dotterels stood up and ran down to a little ridge below.

On 5th June, 1969, Pulliainen watched a pair copulating. Between 09.00–10.15 the hen went to the nest but did not lay. At 10.15 the cock made scrapes about fifty yards away, and the hen went to the nest and sat down 'without showing its cloaca'. The cock then mated with her.

Sexual behaviour after egg-laying

I have occasionally seen dotterels copulate after steady incubation has started. In Finland Pulliainen (1971) also records this.

After dotterels have lost their eggs, perhaps too late in the season for the hen to replace them, the pair sometimes continue to copulate.

On 29th June, 1934, for example, I watched a robbed hen lying as if brooding on a sandy bank. The cock immediately ran up and twice mounted her. She then rose, ran a few yards and again settled into the brooding position. Meanwhile, the cock first preened on a mossy tump, and then for the next seven minutes he splashed and bathed, jerking drops of water over his shoulders. Then back he went to the tump, while the hen now sat on the gravel a few inches away. A few minutes later the two dotterels started to make scrapes on a stony flat, but this led to nothing. The hen laid no more eggs in that summer.

Cocks running with broods sometimes copulate with apparently strange hens. Rittinghaus watched the cock RY, then tending two chicks, tread another dotterel. He also found that cocks with chicks sometimes associated with two other dotterels, violently attacking one, but tolerating the other.

Hen groups

After they have laid their eggs and the cocks are brooding them, hen dotterels form groups. These small parties of grass widows then roam over the hills, spending many hours close to burns or squashy places, and sometimes flying to distant feeding grounds at night. In these groups the hens spar and fight rather half-heartedly, but they sometimes spread their tails and jump over one another. In the Cairngorms and Grampians these parties seldom contain more than half a dozen birds.

Several observers have recorded watching small trips of brightly-coloured dotterels on the move while the cocks were brooding. On 3rd June, 1922, Gilroy came across 'a little band of dotterels, all in remarkably bright and beautiful plumage' and on 10th June, 1940, George Franklin also watched a trip of five large bright hens 'without a care in the world' on the same hill. On Hardanger Vidda, Blair tells me that he watched dotterels make early morning flights, in which trip after trip of four to eight birds flew down beside a swampy tarn. These trips, however, did not consist exclusively of hens as a cock was shot in one of them. On the Dovre Fjell, E. Barth also remarks that 'several hens usually band together and, without responsibilities, roam around in the alpine zone'.

In Holland, in 1963, J. Sollie recorded assemblies of up to thirty hens on the newly-colonised nesting-polders, but some of these were probably non-breeding birds. Only four nests were actually located. The functions and membership of these large Dutch assemblies thus possibly differ from those of the smaller hen groups observed in Scotland and Scandinavia.

During incubation, some cocks periodically associate with groups. For example, Rittinghaus found the cock Black-Green feeding with four other dotterels about one and a half kilometres from the nest on which his hen Green-Green was then sitting. At 03.27, on 12th July, Black-Green was again one of eight dotterels on a different slope of the nesting hill. Rittinghaus, however, was unable to determine the sexes of the other dotterels.

From time to time, particularly in the late afternoon or evening, a hen leaves the group to visit the cock at the nest and then perhaps escorts him to and from the feeding ground. I have even known a group of hens do this.

I can only speculate about the selection value of hen groups. Extensive colour banding is now the most likely source of real knowledge. I believe, however, that these are possible advantages.

1. The collective eyes and ears of a group are more likely to detect predators than those of an individual. The sharpest senses are thus at the service of all. On the other hand, predators are still likely to capture the less mobile birds, thus conserving the strongest and most vigorous.

2. Hen groups exploit niches in the habitat which cocks and broods usually avoid. As I shall later suggest, the dotterel's single parent fledgling pattern helps to conserve scarce food for small chicks.

The flocks frequently assemble and display on particular locations – often on flats close to burns. Are these locations – at which groups of hen dotterels nest and fight – true leks?

A lek is a definite location to which a particular group repeatedly returns to display and wherein each bird dominates a particular part. Judged by this definition, hen dotterels do not establish leks. But group activities in these rather primitive arenas presumably have survival value.

(a) Combats or mock combats harmlessly release aggressive drives. We already know, for example, that in summers producing a surplus of hens some of these 'ladies' mob paired cocks and try to lay their eggs in the nests of mated hens. In 1941 I watched this kind of mobbing on two different territories, and in 1968 A. Tewnion saw a mated hen try to drive off two others, one of which finally laid an egg in her nest.

Well-knit hen groups probably inhibit this kind of anti-social behaviour.

(b) The group may also evolve a rough hierarchy, thus enabling any mateless cock in breeding condition, or one which has lost its mate, to find a vigorous successor. Groups thus provide a social nucleus offering different kinds of services.

Use of colour patterns in display

1. Frontal colours – a chestnut-red breast and black belly patch are used in threat and anger situations. During nest-dances the broad white pectoral band of the hen possibly sexually excites the cock, but she apparently does not employ any special movement to display it beyond half-rotating in the nest.

2. Hens display their grey or silvery underwing surfaces in two different epigamic displays.

(a) They raise their wings high above their backs before chasing and isolating a cock. Paired hens also raise their wings and spread their tails at the end of a sex-chase. Cocks also occasionally make similar play with their underwings. Nervous dotterels, however, often raise their wings and then lower them shortly before flying away. Wing stretching, therefore, probably indicates 'intention to flee', thus suggesting that the displaying bird fears the other.

Dotterels do not violently beat or flap their wings, or goose-step or trip up to their mates as cock greenshanks and redshanks do in pre-copulation dances. A cock dotterel, however, occasionally raises and holds both wings high above his back, just before he mounts the hen or to balance himself during copulation. Ritualised wing stretching in epigamic display is, however, poorly developed.

(b) A hen, with a cock standing behind her, sometimes squats between or beside stones, or in a tussock of grass or moss, and then repeatedly raises, waves and lowers her wings. This flashing of grey silvery underwings probably sexually excites the cock.

3. The cinnamon and creamy-white undertail feathers which the hen shows off in a tilted run in front of the cock, excite sexual desire. Both cock and hen also show off their undertail feathers in nest-dances and egg ceremonies during which they often copulate.

When the hen is squatting on scrapes in tussocks of grass, her broad white superciliary stripes meeting at the nape are conspicuous. Before mounting, the cock sometimes touches or possibly nibbles the hen's nape and upper mantle. The contrasting brown and white head and nape pattern thus possibly stimulate him.

4. (a) Just before attacking a rival, and while approaching it, feathers indrawn, tail and body parallel with the ground, a dotterel suddenly fans its tail to show off the white tips.

(b) In courtship, particularly just after she has chased and almost reached the cock, a hen sometimes half-spreads her tail and simultaneously raises her wings. As with wing-stretching, however, this is not a strongly developed sexual display.

Aggression

A dotterel's frontal colours – chestnut-red breast and black belly patch – possibly assist in dispersion. We have often watched a hen standing erect and upright on tump or hummock, a posture in which she creaks monotonously. This conspicuous stance possibly acts as a distance threat, thus inhibiting the approach of rivals.

2. Before running towards a rival, a dotterel often stands upright, puffing out its breast and showing off the black belly patch. This posture suggests clash between aggression and a desire to flee.

3. Attacking dotterels approach rivals in stops and starts, running with short quick vibratory steps, necks indrawn, and bill and back horizontal to the ground. They now look almost like small balls rolling along the ground. Just before attacking, the angry dotterel ruffles its back feathers and often half-opens its tail. Both cocks and hens use these displays although fighting hens most frequently employ them.

In territorial fights, however, which are really double battles – hen against hen and cock against cock – both sexes behave like this. A pursued or dominated hen starts to run away in horizontal posture, but soon becomes upright and erect. Periodically it turns laterally and occasionally tilts forward, dropping one wing and fanning its tail almost vertically. This posture seems to inhibit attack.

Subtle nuances indicate degrees of dominance and subordination.

4. Trilling shrilly, and moving as if on rails, dotterels sometimes raise or half-raise their wings immediately before leap-frogging over one another. They sometimes actually strike rivals, plucking out feathers as they pass over them. The two battlers then immediately turn and face one another.

5. A dotterel occasionally flies straight at a rival without preliminary display, striking its enemy with beak or wings. We have watched particularly fierce fights between rival cocks on the borders of territory.

6. Seton Gordon tells me that he once met with a hen which dived at his head, almost like a peregrine.

Distraction displays and behaviour

Dotterels have a battery of injury feigning or distraction displays which they employ to lure or deflect predators away from eggs or chicks.

1. The dotterel runs fast, tail spread and depressed, wings partly raised, giving the bird the appearance of having a hump. In this display the white outer-tail feathers are conspicuous. Some of the few recorded brooding hens use or have used low intensity forms of this display.

Ernest Blezard (1926) beautifully describes the way of a dotterel in the English Pennines. 'It gradually walked further and further away in a series of half circles; finding no response on my part, it ran back to my feet and repeated the performance. These manœuvres were carried out rapidly. The bird creeping with legs well bent and breast close to the ground, the broadly striped head was retracted in line with the back and tail depressed, fully spread, and trailed along the ground. The rump feathers raised and wings held loosely. The yellow margins of the ruffed upper plumage and the plucked out chestnut flank feathers attracted the eye. Each time the bird described a curve, the outside wing was momentarily raised.'

At other times dotterels repeatedly raise one wing as a variation to raising both.

2. The dotterel runs with drooped and beating wings, but closed tail.

3. It again runs with drooped and beating wings but with its tail partly open and dragged along the ground.

4. It lies flat on the ground, with legs trailed and scratching, or lies passively on a stone, squeaking piteously and spreading its tail like a fan.

5. It does the 'mammal run' – fast running, all feathers indrawn, and low crouching stance.

6. 'Mock-scrape making' – the cock scrapes and half-rotates and throws moss or grass over his shoulders. In my experience, an unusual display. In July, 1945, I watched a cock with a nine-day-old chick do this.

7. 'Mock brooding.' I once saw a cock squat in a tussock as if brooding, and then drop its head and appear to arrange imaginary eggs. It then had hard set eggs in the nest. An inner conflict between two conflicting drives – to brood or to flee – probably caused this.

8. The cock flaps his wings high over his back. Another cock continuously flicks his wings upwards, opening and shutting them.

9. I do not know how to classify the display of a cock which 'explodes' from the nest to deflect what it regards as a large, but rather harmless, mammal. On 16th June, 1934, for example, the little cock, Ramsay, repeatedly did this and also stepped off his eggs and ran towards us in upright posture while flapping his wings. On 10th June, 1969, Derek Ratcliffe had the same experience in Lakeland. 'When I put my hand out to touch the bird it suddenly erupted vertically off the eggs to about 2–3 feet and pitched a couple of feet to one side, screeching angrily at me with a *peer-peer-pee* call. It very soon walked back to the three eggs and settled down on them again. After taking more pictures while lying beside the bird, I again tried to touch it. This time it not only erupted, but flew at and touched my head before pitching beside the nest.' While these displays are better considered as aggressive, brooding lapwings often use this behaviour to deflect sheep or cattle which are about to tread on their nests.

10. During distraction displays or when running about close to a nest, a dotterel pecks downwards without necessarily eating anything or even touching the ground with his beak. This is presumably 'displacement feeding' or just possibly a 're-directed attack'.

11. Distraction flight. Slow, heavy, wing-flapping, with partly expanded and depressed tail and dangling legs. This resembles one of the golden plover's distraction flights. Both cocks and hens sometimes fly like this. My records of hens are of those disturbed at the nest after laying and then brooding an egg.

CHAPTER 5

Petticoat Government

FOR long ornithologists did not realise that dotterels had reversed sexual roles and plumage differences. But E. T. Booth had flashes of intuition and inspiration when he outshone his contemporaries. In 1866, while skinning his two trophies in Glen Lyon, he discovered that the bright bird procured from a small trip was 'a female, while the bird shot with the brood turned out, contrary to my expectations, a male. The brightness of the colouring on the feathers of all the birds composing the small parties on Cairngorm would lead one to believe that these were females, the care of the juveniles appearing to devolve on the males who at this season exhibit more worn and far less showy plumage.'

For many years on, however, most British ornithologists remained sceptical or unconvinced. In the early 1920s, however, by watching their mating, Carrie satisfied herself that the hen was larger and brighter and that the drabber cock normally sat. Continental observers have been able to distinguish between cock and hen. H. Franke (1952) found that the cock was smaller, duller in colour, and had many flecked feathers. The hen had a completely black or dark brown crown, a brilliant white stripe above her eyes, a clear white crescent above the breast, a large black belly-patch, and a brighter and greyer mantle. Franke only found cocks brooding eggs.

On the other hand, Rittinghaus was less definite. The hen, however, had a darker crown, a smoother grey-brown mantle, a bigger patch of black on the belly, and was marginally larger and more colourful. But these distinctions only seemed clear when the two dotterels were at the nest, standing together or 'in the hand'. Rittinghaus proved that some hens brooded.

My own observations are clear-cut. I have now seen five hens lay eggs. Every time the hen was larger, brighter and plumper than her mate. I have also watched the mating of eight pairs; the duller bird invariably mounted the brighter; I never saw reversed mounting. I also regard the darker-brown crown, the apparently broader and purer white head-and breast-bands, the larger head and thicker neck, and general puffiness as useful field marks for hens. In Scotland the cock predominates in parental rôles.

Discussion about waders with reversed roles

Few birds have evolved reversed sexual rôles. How then did this pattern emerge and why has it survived?

All forms of reversal are unlikely to have had a common origin. Let me discuss the possible origin and pattern of reversal in the dotterel. K. H. Voous (1962) speculates that the dotterel's distributional history is possibly linked with that of the caspian plover and that its highly fragmented breeding range suggests that it is an arcto-alpine relict. Although the dotterel is now a mountain bird in many locations, he suggests that it was originally a tundra species, the change probably having occurred during the Pleistocene period. Hans Johansen believes, on the contrary, that the dotterel, originally a mountain bird, has only recently colonised the Arctic. These contrary speculations show that we can do little better than make informed guesses about the dotterel's early distributional history.

Our bird's ecological niches in both tundra and mountain habitats are probably significant. In the tundras of U.S.S.R., for example, dotterels avoid lowland swamps covered with dense grass or shrub; and usually breed on stony hills with carpets of short grass, lichens and Iceland moss.

In equally specialised montane biotopes they normally nest on the slopes or ridges of broad-backed hills or on mountain plateaux and tablelands with stoneshot carpets of short grass or moss. They generally avoid angular, pointed mountains, or unbroken scree fields profusely littered with large boulders. They thus select broadly similar biotopes in tundras and in montane zones. Although nesting close to arctic birds, like snow bunting and ptarmigan, the exquisite dotterel always seems rather alien, although its colour-pattern is remarkably cryptic in arctic uplands.

In Central Asia, the caspian plover *Charadrius asiaticus* is a bird of grassy steppes, inland desert saltings, and mountain tablelands with carpets of short grass. Little is known about its behaviour and breeding biology, but both sexes apparently sit on three eggs and both have brood-patches.

The dotterel's affinities with the plovers of North America are uncertain, but Voous suggests that its distributional history may be linked with the mountain plover *P. montanus*, a bird adapted to almost desert environments in dry stony grasslands in western U.S.A.

W. Graul (*in litt*) finds no evidence of reversed courtship in a large population in Colorado, the more aggressive and slightly brighter male doing the earlier courting. Mountain plovers form monogamous pairs but only one member in each pair incubates and tends the chicks. A hen drove the cock from the nest after she had completed the clutch. As soon as the hen has 'laid out' the pair-bond dissolves. In their almost waterless desert habitats the mountain plover has thus evolved different breeding adaptations from those of the dotterel. Both species, however, produce clutches of three not dissimilar eggs.

The distributional link between dotterel and caspian plover seems clearer. In some glacial phase the common ancestor, a bird of dry grassland steppes, was probably forced to retreat through loss of suitable habitat. Then, in a favourable period of climate, it possibly failed to compete with other species. The specialised short-billed plover was then probably forced to occupy less favourable moss-grass habitats in the tundra and subsequently on the slopes of rounded fells and mountains. Lacking serious competition there the emerging dotterel thus successfully exploited these niches. In high montane grasslands in North America, on the contrary, a dotterel species possibly failed to compete or to adapt and thus became extinct. The recent colonisation of unlikely habitats in Holland possibly echoes early distributional struggles in primeval steppes.

At first both sexes probably shared parental duties and were roughly of equal size and brightness. However, the short breeding season in demanding biotopes – and heavy snowcover, low summer temperatures, or periods of drought – could have enforced readjustment. In these conditions, the hen possibly could not produce and perhaps replace eggs and simultaneously take an equal share in parental duties. A new pattern thus gradually emerged, the hen concentrating on status, selecting a mate, winning and holding territory, and in producing three eggs at daily intervals. The cock, meanwhile, helped to secure and defend the territory and mainly brooded the eggs and reared the young. It was thus an advantage for the hen to produce eggs for the fresher cock to incubate. The combination of egg-laying and brooding might have adversely affected her health and vigour. As we shall later see, a scarcity of food for the chicks was another likely factor.

In the new régime, however, males were at greater risk. That being so, selection would ensure a small surplus. I have no precise data on sex-ratio in dotterel, but in twenty-two Wilson's phalaropes chicks there were thirteen cocks and nine hens (Höhn) – possibly a suggestive trend.

A hen dotterel's new rôle, on the contrary, now demanded different characters. Larger size and brighter colours would assist the partner which had to dominate, woo, win and hold. A larger bird could also be better adapted to produce eggs quickly. Freedom from parental duties also might assist a hen to replace lost eggs or to lay a second clutch for a spare mate. To the dotterel, whose hens sometimes produce two complete clutches within eleven days, polyandry had something to offer. Reversed sexual rôles clearly favoured this pattern.

Recent physiological research seems to confirm these speculations. J. E. Johns (1964) proved that injections of male hormone (testosterone) induced cock and hen phalaropes to grow new and bright nuptial feathers to replace those plucked out of their winter or juvenile plumage. E. O. Höhn and S. C. Cheng (1967) also discovered that a hen Wilson's phalarope's ovary sometimes contains more male hormones than a cock's testes; this probably explains her greater aggression and brighter plumage.

A combination of prolactin and testosterone is needed to produce brood-patches. By injecting testosterone and prolactin, J. E. Johns and E. W. Pfeiffer (1963) induced out-of-season brood-patches in red-necked and Wilson's phalaropes. Höhn and Cheng (1965) later discovered that although cock phalaropes produced prolactin, hens produced less or even none. This possibly explains the absence of brood-patches in hen phalaropes. Höhn speculates, however, that both sexes originally brooded eggs and shared similar androgen-dependence, but that in a later phase of evolution 'a hereditary deficiency of prolactin secretion appeared in the females. Thereafter, only males produced brood-patches and only they incubated . . . Selection would thereafter favour dull plumaged males, i.e. those which at the time of the formation of the nuptial plumages were relatively poor androgen secretors. In females, on the other hand, once they had ceased to incubate, there was no restraint on androgen production, which eventually came to exceed that of the male.'

Hen dotterels take a greater share in parental duties than phalaropes. This possibly led to less distinctive dimorphism and development of brood-patches by some hens (H. Rittinghaus).

The dotterel's fledging pattern is abnormal. If the hen attempts to join young chicks, the cock often drives her away. Later on, however, when the chicks are more mobile, he sometimes allows her to join.

Why this puzzling behaviour? By defending the nesting territory during incubation, the cock ensures that the chicks will 'inherit' an untapped food supply while at their weakest and least mobile. He also now prevents the hen sharing this family food reserve.

Ill-adapted to attack predators, dotterels also largely depend on elaborate displays to distract predators. The chicks have cryptic colour-patterns and soon learn to conceal themselves. In this phase, therefore, the more conspicuous hens would attract predators and consume scarce food. The hens, meanwhile, stay in small groups and often feed on ground which family parties avoid. Later, when the chicks are stronger, the cock often accepts the hen. This is not, however, an inflexible rule. In the Cairngorms, for example, some hens leave the nesting grounds ahead of cocks and juveniles. In east Siberia some hen trips also leave in the second half of July, whereas most males and juveniles stay there until the end of August (Ivanov, *in litt*). The dotterel's patterns thus resemble those of other arctic waders with single-parent systems. Similar pressures are possibly responsible.

The dotterel was originally a species of the montane steppe and apparently widely distributed on lowland steppes and tundras in later glacial phases. Its summer life, therefore, is now a stern struggle in fundamentally alien habitats.

Length of nesting season in relation to ecological needs

In summer, dotterels and their broods largely feed on soft insects – particularly *Diptera* – *Tipulidae* and their larvae, and spiders. The availability of this food largely determines the commencement of an invariably brief nesting season. In the Cairngorms, for example, weather and snowcover in the second and third weeks in May are often decisive. In 1951, a year with heavy snowcover on higher plateaux, few dotterels started to lay until the last week in June – over six weeks later than the earliest hen laid her first egg on the same ridges in the hot summer of 1940. In Fenno-Scandia and U.S.S.R. the breeding season is almost invariably shorter. In late summers some pairs possibly fail to nest on some of their breeding fells far north of the Arctic circle. In Finnmark, for example, Trafford noticed purple sandpipers were nesting on snow-covered hills while dotterels were still in flocks. But we still lack proof that dotterels do not breed in backward summers. In 1951 and 1968, for example, some pairs finally reared young which could not fly in mid-August. I doubt, however, whether dotterels could successfully rear young as late in really backward Arctic summers, where the breeding season is much shorter.

Apart from time spent in pairing, courtship, territorial settlement and nest-dances, dotterels require approximately eight weeks to lay, brood, and hatch their eggs and rear their chicks to free-flying. In July, moreover, adults begin to moult before they leave the nesting grounds.

Table 1 gives comparative data for other boreal and arctic waders. Dotterels take from fifty-two to sixty-five days to produce fledged young, whereas ruddy turnstones need only about forty-four to forty-six days. The smaller Baird's sandpiper requires forty-five to forty-seven days, and the much smaller grey phalarope thirty-eight to forty-four days. The larger American golden plover also produces fledged young in about fifty-five days. On the other hand, the mountain plover, a species adapted to an exacting environment, apparently requires about sixty-two to sixty-five days from first egg to fledged young (Graul, *in litt*), and the sanderling up to fifty-four days.

This certainly suggests that particularly 'difficult' habitats demand special adaptations. Reversed roles have thus probably helped dotterel and phalaropes to endure.

CHAPTER 6

Polyandry

GEORGE BLACKWOOD (1920) suspected that dotterels were sometimes polyandrous. At a nest in June, 1907, there were three 'obviously
alarmed' dotterels. 'All three birds showed anxiety only when we were
in the vicinity of the nest.'

I have never watched dotterels behaving in this way, but two ornithologists have seen extra birds at nests. In Finnmark George Trafford
found an extra bird present each morning and twice noticed that one
had an injured leg while the other had not. If the two birds *were* cocks,
was Trafford watching an unusual form of polyandry? On Skokholm,
for example, M. P. Harris (1967) watched a banded cock oystercatcher
intermittently brooding on two nests a hundred yards apart! He apparently belonged to the second nest and was an interloper at the first.
'Another male and female were caught on the first nest.' So the almost
incredible does sometimes happen.

In 1970 Jack Robson (*in litt*) watched a pair of dotterels 'collect' a
third bird which accompanied them to their nesting area. 'I would guess
this third bird was a hen as two seemed slightly larger than the third.
I had a similar experience in 1955 when a pair of dotterels would not
show me the nest, but went off and collected a spare hen. Eventually I
found the male of the second hen sitting, but I never made any sense of
the original pair.' Here there were trios but no suggestion of polyandry.

But a few hens really are polyandrous. On 27th May, 1934, Carrie
and I watched a large and extremely bright hen running with two cocks.
The three dotterels often returned to a rounded slope. The hen was
broody but the cocks sparred with one another. We were particularly
interested in this hill flat as Blackie, a hen laying almost unique eggs,
had nested there in 1933. On 10th June, while hunting the hill where we
had seen the trio, I saw a large hen on the skyline. Then a second
dotterel flew low and pitched on the slope above. Ten minutes later I
saw it fly 100 yards uphill and drop down beside a triangular green stone
from which I flushed it from three magnificent eggs. Only Blackie
could have laid them! As I was looking at them, a hen suddenly ran
along a ridge about twenty yards away and then another dotterel called,
and the hen flipped away. Blackie's eggs were slightly set. She must have
'laid out' about 3rd or 4th June.

On the same afternoon, and again later in the evening, we watched two dotterels working at a scrape about 200 yards SE of Blackie's nest, but they never laid in it.

On 25th June we watched two dotterels change duties at a nest and flushed the *hen* from three unmistakable eggs. Blackie's again. But these eggs were incubated about ten to twelve days. Now the truth was out! Blackie had completed a second clutch about 14th to 15th June, precisely when we had expected to find eggs in the deserted scrape. She had thus 'put down' a new mate about eleven days after leaving the first cock brooding her first clutch.

We have watched other trios but have never again had real proof. In June 1938 I saw a hen running with two cocks. I later found one nest on that ridge, but if there was a second I missed it. In 1940 we watched a pair run towards a cock who was sitting on deeply incubated eggs. The hen ran almost up to the nest, but the brooding cock rose, ran past her, and drove off the rival cock. Then he returned to brood. In the afternoon this pair nest-danced and copulated about 150 yards away. A fortnight later we found *their* nest with long, pointed and well-blotched eggs, remarkably similar to those in the first nest and quite unlike those of any other dotterel found in 1940.

In both 1949 and 1961 we saw hens running, flying and moving with two cocks, but we never proved that they produced clutches for both mates.

On 30th June, 1947, Adam Watson caught and ringed two dotterel chicks which were just beginning to flutter. During the distraction antics of the drab male dotterel, a large bright hen ran up, but did not display. On 4th July, at the same spot, Watson raised a much brighter cock with two chicks about one or two days old. Again the cock injury-feigned and a large bright hen briefly joined them. He later found the first brood close to a burn about two hundred yards away; these youngsters were old enough to flutter a yard or so. Emphasising how seldom the hen arrives when a brood is disturbed, Watson suggests she was possibly interested in both broods. The ages of the broods suggest polyandry.

Dotterels are also possibly sometimes polyandrous in England. On 29th May, 1969, R. Laidler found a nest with three pale speckly eggs on an English fell. On 10th June D. Ratcliffe found a second clutch about two hundred yards away, but 'the best marked egg had been dented and the albumen had congealed over the dent.' On 14th June, however, G. Horne and R. H. Brown found the second nest deserted, and the previously dented egg lying sucked a few feet away. The two deserted eggs had been incubated about eight to nine days. 'The proximity of the two nests and the difference in laying dates – at least 11–12 days – make it possible that the two clutches were laid by a polyandrous hen.

1. Dotterel nesting grounds in the Cairngorms. *Above:* Snow in May can delay nesting, and summer snowstorms may cause brooding dotterels to desert their eggs, although they often sit through them. *Below:* Grassy tussocks on gravel terrain.

2. *Above:* A thin spidery cock dotterel standing alert on a rock in Sweden. *Below:* A cock dotterel about to brood his eggs in the East Grampians. He often swivels and pokes downwards before settling. Note the woolly fringe moss carpet.

Laidler's photograph showed a spotted egg which could have been laid by the same bird that laid mine. They were certainly not sufficiently dissimilar to rule out this possibility. The peculiar failure of the one nest could in any case be due to interaction between the two males; it would be unlikely to have been caused by a predator which would presumably have taken all the eggs.' (Ratcliffe, *in litt.*)

Polyandry abroad

AUSTRIA

In May, 1952, Hans Franke watched a triangle in the Austrian Alps. A hen, identified by an irregular pattern of feathers forming a deep V or Y under her pectoral band, ran the hill with two cocks, one almost as bright as she was, the other much duller. Time after time the duller cock attacked the brighter one and sometimes drove him away. On 1st June, however, the bright cock was still close to the nest which now had one egg in it. But the drab master cock was also still there with the hen, and at 5.05 on 3rd June it was the dull one which shuffled off the eggs. Hardly had he done so, however, before the hen went on to the nest. Between 05.15 and 06.30 she laid her third egg and within a few minutes had cleared off. Next day Franke only had a brief glimpse of the hen when she circled around the nest several times. The dull cock then brooded the three eggs from which two chicks hatched on 26th June. Then, on 8th July, the beautiful hen was back on the territory and with a court of two cocks, one decidedly brighter than the other. Once more she mated with the duller-coloured suitor, completing her second clutch on 15th July. The polyandrous hen and the bright cock then disappeared, leaving sobersides to brood!

Franke explains how he recognised this hen.

1. She had a peculiar dark V or Y under her pectoral crescent.

2. The last egg in both clutches was sparsely marked, rough-shelled, and infertile.

3. The hen selected both scrapes; the second was only about twenty yards from the first.

FENNO-SCANDIA

Rittinghaus watched several cocks with chicks associating with hens. Cock RY, for example, copulated with one attendant hen, but he never came across an unexpected clutch in any territory.

Professor W. Hobson watched a hen in Finnmark which possibly had three mates. On a hill of about a hundred and fifty acres – the only dotterel habitat in many square miles of forest and marsh – Hobson found three clutches at different states of incubation. The only hen there escorted a cock right up to a nest with eggs set about a fortnight. She also advertised over both the other nests. All three nests contained eggs of the same type.

Pulliainen (1971) made another striking discovery. After laying three eggs in one nest a colour-marked hen laid the first of a second trio of eggs 4.7 days later. The distance between the two nests was 600 metres. This was another remarkable hen. Like Blackie, she shared the incubation of the second clutch with her new mate. My delight and excitement at the Austrian and Finnish discoveries can be imagined.

Polygamy in dotterels

Polyandry is likely among groups with more cocks than hens. Few groups, however, are of this pattern. In 1941, 1942 and 1968, for example, there were more hens than cocks in some Cairngorm populations. In 1941, strange hens mobbed and chased brooding cocks when they left their eggs. On 5th June, 1941, moreover, Carrie found two nests less than a hundred yards apart. At the first a cock was brooding three eggs, a hen two eggs in the other. We watched the second nest, but never saw a cock near it; the eggs were infertile. Had this second hen joined a mated pair and then produced two unfertilised eggs?

On 15th June, 1968, Sandy Tewnion came upon three dotterels fighting and bickering. Two hens ran a few yards apart, 'giving the distraction call of the female when disturbed at the nest'. Then a third hen joined them and started to run about as if trying to lure him from a nest. After a long search Tewnion found a cock sitting on one egg to which it soon returned. To his surprise, one hen went to the nest and the two others also ran up. The first hen now chased the two others away. One of these hens, however, kept running almost up to the nest. 'No matter how much the dominant hen chased her away she did not retreat further than about 50 yards from the nest. The third hen, on the other hand, did not approach so closely, but was content to stay about 50–60 yards away.' Tewnion photographed the single egg. On 29th June, however, he found the cock sitting on four eggs: three clearly laid by one hen and the fourth by another. The photographs, and later the hatched egg-shells, finally proved that the dominant hen had laid the egg found on 15th June and that a second hen had later managed to lay a single egg in the scoop. Presumably, the dominant hen finally expelled her rival, which possibly laid out elsewhere. Nevertheless, the interloper's egg hatched out. This cock or another had fertilised her.

Lesbianism

In 1953 G. Trafford flushed a hen dotterel from six eggs in Norwegian Lapland. After watching for two days he discovered the nest was shared by two hens, both of which he collected. 'The first hen sat very closely and feigned injury when flushed. The eggs were infertile and without traces of incubation. If the two hens had been mated to a cock, some of the eggs would presumably have been fertile.' All other dotterel groups on surrounding fells already had broods. Reg Wagstaffe, of Liverpool Museum, kindly examined the skins of these hens, but could find no brood-spots on them. Were these two hens, then, actually Lesbian mates? Lesbianism seems more likely than that two hens had shared a cock which had failed to fertilise them.

Is it possible that dotterel and phalaropes have evolved extremely flexible mating systems with polyandry or polygamy as expedients to cope with a temporarily unbalanced sex-ratio? As cocks and hens tend to arrive separately, disaster or heavy losses on migration could affect the numbers of either sex. In species with reversed sexual rôles, however, polygamy is of improbable selective advantage, as it would impose upon the cock double the normal complement of chicks. The mobbing of mated cocks in years with a surplus of hens is a bad bet selection-wise. On the other hand, polyandry enables a hen to concentrate on status and egg production and allows unmated cocks to brood and hatch extra clutches.

Polyandry in other birds

A few other birds – usually those with reversed sexual differences – are occasionally polyandrous. Phalaropes have reversed sexual rôles and colour differences but, like the dotterel, are apparently only occasionally polyandrous. There are, however, a few records. In Alberta T. E. Randall (*pers. comm.*) discovered that several hen Wilson's phalaropes had mated with two cocks. 'The outstanding case took place close to Peace River, Alberta . . . On extensive lake meadows I found a female and two males – the only phalaropes in this particular area, and eventually flushed the males from nests, each containing 4 eggs. These nests were 100 yards apart. One clutch hatched 11 days later than the other.'

Randall also proved that Wilson's phalaropes were able to produce replacement clutches. Twelve days after taking clutches of fresh or slightly incubated eggs from an isolated breeding colony, he found three replacement clutches; 'but only three eggs were eventually laid in each of these nests.'

Red-necked phalaropes are also occasionally polyandrous. L. W. Montgomery watched a trio of red-necked phalaropes in Co. Mayo and later discovered two clutches of almost identical eggs in nests about ten yards apart. The same hen had clearly laid them all. On the day that Montgomery found them one set was fresh, the second slightly incubated.

The hen red-necked phalarope, which Tinbergen (1959) watched in east Greenland, commenced a new cycle after the cock had started to brood her four eggs, but she did not succeed in attracting a second mate. Tinbergen suggests that 'successive polyandry certainly seems possible in this species.'

Polygamy in phalaropes

Phalaropes are also sometimes polygamous. In May, 1921, Randall found eight Wilson's phalaropes' nests on a small grassy island. Five contained four eggs, two held five, and the eighth six eggs. The extra eggs in the two larger clutches differed from the rest of the set, but were identical in shape and colour-pattern with each other. 'This led me to make a closer examination of the group, and I eventually found that this colony consisted of 8 males and 9 females.'

Cock red-necked phalaropes are also occasionally bigamous. In 1929 W. M. Congreve and S. W. P. Freme (1930) recorded more hens than cocks in some Iceland nesting groups. One cock was seen copulating with two hens, one after the other. Later, Congreve found two nests each with eight eggs. In phalaropes, however, polygamy is likely to be rare, casual, and selected against.

Possible origins of polyandry

To the best of my knowledge, dotterel and the phalaropes are now normally monogamous. But, in earlier evolutionary phases, were all or any formerly frequent or regular polyandrists? I have conjectured that at one stage of evolution selection pressures probably produced a male surplus, to cope with concentration on parental duties. In summers when males outnumbered females, some would tend to mate with 'grass widows'. Although possibly producing fewer flying young than those reared from earlier first clutches, these second broods would help to compensate for the deficiencies of a single-parent pattern.

Casual polyandry in passerines

A few normally monogamous passerine species are occasionally polyandrous.

Some nesting groups of starlings, for example, contain more cocks than hens. Both sexes normally incubate, but if a particular hen has a weak brooding drive, she sometimes accepts a second mate and lays a fresh clutch for him. In other colonies, however, some cocks are polygamous, mating with several hens in succession. Starlings thus also tailor their mating systems to current sex-ratios.

Polyandry in other waders

Dotterel and phalaropes are not the only polyandrous waders. In India and south-east Asia, the hen pheasant-tailed jacana or Chinese water-pheasant, *Hydrophasianus chirurgus*, defends a territory where she sometimes courts several cocks in succession, then lays three or four clutches at intervals of nine to twelve days. Each cock, in turn, then broods a clutch of eggs and rears the chicks. If the eggs are destroyed the hen provides the cock with a replacement clutch.

In the Old World painted snipe, *Rostratula benghalensis*, the larger, brighter hen holds a territory where she courts a cock and then leaves him to undertake all parental duties. Thereafter she often takes a second mate, sometimes completing a new clutch before her first has hatched. In one nesting group there were only four hens to twelve cocks. Like dotterels, hen painted snipe also gather in small grass widow groups. These birds breed almost throughout the year in favourable habitats.

Tinamous

The neo-tropical tinamous, birds closely related to rheas, have evolved different forms of polyandry and polygamy. Hen variegated tinamous, *Crypturellus variegatus*, often mate with up to four cocks. Each hen lays a single egg in the cock's nest-scoop, leaving him to brood it and to rear the young. A cock sometimes broods three eggs, one after the other, but the hen may lay her next egg while the cock is still rearing a well-grown chick.

D. A. Lancaster (1964) describes successive polyandry and harem polygamy in the Boucard tinamous, *Crypturellus boucardi*, and in brushland tinamous, *Nothopracta cinarescens*. In these, the cocks establish territories and then mate with two to four hens which all lay in a common nest. The cock then hatches the eggs and rears the young.

Button quails

D. Seth-Smith (1905) studied reversed courtship in button quails, *Turnicidae*, birds of tropical and sub-tropical grasslands. In *Turnix tanki*, the larger and more dominant hen possesses a bold red collar which the smaller cock lacks. The hen lays three eggs, leaving the cock to brood them. While he is sitting on a nest or guarding and rearing the brood, she seldom visits him.

The hen Australian button quail, *Turnix varia*, booms loudly on her territory where, in the cock's presence, she makes a nest-scrape. If the cock fails to accept her the hen resumes booming. During courtship she gives the cock 'dainty morsels . . . calling him with a faint clucking sound, in the same way as a rooster calls to his hens.' The cock then broods the eggs and the hen booms to call up another mate.

Prince K. H. Dharmakumarsinghi (1945) recorded reversed courtship in the bustard quail, *Turnix suscitator taigoori*, which has evolved a slightly different pattern. The hen sometimes visits the nest and broods the eggs while the cock is feeding. Probably, however, she does not really incubate them. While the cock is sitting the hen often booms, but after the hatch she joins cock and brood. The bustard quail's breeding system is thus similar to that of some of our dotterels.

Hen button quails, *Turnix sylvatica*, also produce clutches of eggs for several different cocks. Each time that the cock is down the hen booms to attract a new mate. The *Pedionomidae*, a group closely related to button quails, are also possibly polyandrous. Here also the cocks brood.

Tasmanian Gallinule

About half hen Tasmanian gallinules apparently form permanent pairs with two or three cocks, which help to rear the young. They thus have a selective advantage over monogamous hens (M. G. Ridpath, 1964).

Red-legged partridge

Red-legged partridges are other birds with unusual reproductive adaptations. D. Goodwin (1953) proved that a particular pair simultaneously reared two broods; the hen laid two clutches, on one of which she sat, while the cock hatched out the other. D. Jenkins, (1957) also found that one hen paired with two cocks and possibly produced three clutches, one of which she herself probably brooded.

All these fascinating reproductive adaptations are difficult to explain. Is the process of natural selection really so rigid that it allows the species only one option? Or does chance sometimes help to beget the un-orthodox?

Take the waders: ruff, buff-breasted sandpiper and great snipe all have breeding systems based on lek displays, but only the male ruff has evolved brilliant and individually characteristic nuptial adornments. The courts and arenas of all three species, although functionally similar, also differ in detail and pattern.

Dotterel and the phalaropes have reversed courtships and sexual differences, now usually allied to monogamy. Painted snipe and pheasant-tailed jacanas, with equally distinctive reversal, practise successive polyandry.

Were all these strangely different adaptations inevitable? Surely other patterns might have been equally successful. To some species, poly-andry now offers the means to exploit abundance or to redress sexual imbalance; to others, bare survival in the jungle of ecological com-petition.

CHAPTER 7

Territory

Establishment and formation of territories

IN Scotland dotterels usually pair from mating groups. Orthodox territories – firmly defended display-centres – thus have little significance in this early phase. Nevertheless, as I have already described, an early lone hen sometimes creaks rustily on a ridge, winnow-glides over the tops, and runs, stops and squats on a future nesting fell. Her displays are, however, brief and spasmodic, and the advertising flights lack the set form and abandon of greenshanks and golden plover, but she flies above vast stretches of tundra or tableland.

Early trips of hens likewise fail to establish firm territories, but they bicker, display, and possibly roughly settle status in mobile rings on flats and hilltops. Calling noisily, sometimes sparring on the ground or jostling one another in the air, they travel widely. Later, dotterels of both sexes belong to these courting groups in which the hens contend for mates. Sometimes two couples – two hens and two cocks, still only loosely paired – fly high, purring angrily and milling on the ground and in the air. But, at this stage, the dotterels are still fighting for mates rather than for territory or living-space. You hear the angry birds reeling well before you see them. Then, their dove-grey underwings flashing in the spring sunshine, they come swishing downwards. Almost immediately after they land the couples first separate and then converge. If one pair does not instantly 'concede' the more aggressive couple runs forward. Soon fighting begins – usually but not always hen against hen and cock against cock. To and fro they glide and slip apart like skilful boxers sparring for openings. Milling intermittently, or sometimes almost continuously like this, one pair usually soon prevails.

The sex-beat now quickens. A few minutes after the defeated couple has fled the victors sometimes start nest-dances or even enjoy one another. On 7th June, 1934, exactly four minutes after the vanquished had departed, Meeson and Sheila were nest-dancing.

These early combats help to disperse the pairs. But a defeated but persistent pair sometimes finally settles and nests on a nearby flat or shoulder. Or, conceding defeat, it shifts over a mile before settling down in unoccupied living-space. For instance, the pair which Meeson and Sheila overcame laid their eggs on a flat over a mile away and higher up the hill. This scattering, however, is most usual when a few pairs are competing for many hundreds of acres of good dotterel ground. On the other hand, in years of heavy snowcover and late thaw, when several pairs are contending for fewer snow-free patches, defended territories are often more compact and nests much closer together. A ridge or hill shoulder, however, often separates two territories, thus reducing fighting, as dotterels react more violently to rivals on the ground than to those flying overhead.

Possessed of a stable territory, the two dotterels often become noticeably less restless. Previously they had probed, explored, contended and joined in nest-dances. Now the two birds work purposefully, often shaping, enlarging, and lining particular nest-scrapes. Then, shortly after she lays her first egg, the hen persistently runs up to the scoop in which she intends to lay, thus showing it to her mate. How broody and passive the dotterels now are! Standing up beside the nest the two beautiful birds often look like stuffed specimens in a glass showcase.

The territory now has real meaning. It provides the dotterels with space of their own – space in which to court, mate and adjust in peace, and without the compulsion to fight and evict. The hen can now conserve the vital energy needed to produce three eggs on consecutive days and the cock's brooding drive matures and strengthens.

Hen dotterels which have failed to mate usually re-join or stay members of trips or groups. Occasionally, however, seeking a second mate after the first is brooding, a hen makes advertising flights far and wide over the hills, but she does not base them on a firm and defended territory. Franke records, however, that the polyandrous Austrian hen put her second mate down on eggs within twenty metres of the nest-scrape in which her first eggs had hatched.

Territorial behaviour during incubation

Is territory still vital after the cock has started to brood? During incubation hens seldom visit mates or nests, usually leaving the cocks to brood the eggs, while they stay in small trips. Cocks, however, often leave their nests to drive off strange dotterels. Purring furiously, they buzz at trespassers, sometimes dropping to the ground, then running a few yards before rising and flying again. Threat alone usually suffices. Trespassers seldom stay to fight. Flying low, following the contours of the hill, they depart at speed. Then – perhaps before the tenant cock returns to his eggs – he characteristically jerks and bobs his head, false-feeds, preens, or scratches himself. His frustrated anger drive thus sparks over. With equal fury cocks drive away trespassing cocks, hens, pairs, or trips. On 13th June, 1936, for example, I watched a cock leave his eggs to chase a wandering hen. Half fanning his tail, like an angry and aggressive ring plover, he dashed at the hen and drove her off. On 1st June, 1961, Brock also saw a brooding cock pursue a pair which had approached his nest. This chase seemed to upset his brooding rhythm. For over an hour he stayed away from his three fresh eggs.

Although flexible and irregular in size, a dotterel's nesting territory does not entirely compress. In years with heavy snowcover and high breeding numbers, two pairs sometimes nest quite close together on the same ridge or hill shoulder. This means fights between the cocks from adjacent nests. The nearer the trespasser comes towards his nest and eggs, the more violently the home cock attacks. Hens also occasionally defend nesting territories, but they usually only meet trespassers by chance. In 1941, however, when there were more hens than cocks, a small group of hens often invaded occupied territories. On 14th June I watched the group run across a ridge where a cock was brooding on freshly laid eggs. Suddenly a great commotion; one of the hens had started upon her companions. Running up and down with rapid, oscillating vibratory steps, bending low, humping her body, and half-unfolding her tail, she seemed almost to rattle with anger. At first the other hens pecked at the ground and sleeked their feathers. Then the angry hen attacked and sent them flying away over the hill. Now she waited until the cock left its nest and flew away with her. Almost certainly she was his mate.

In early nesting seasons a trip of courting dotterels sometimes attach themselves to a brooding cock. The cock leaves his eggs to feed, meets the trip, and they all follow him to the nest. The strong social impulse in dotterels probably explains this strange behaviour. In the third week of May, 1959, Tom Weir watched a party of cocks and hens fly in from afar and 'start a sparring match, jumping in the air, pecking, running, stopping and starting'. Between rounds, a cock suddenly left the group and flew about a hundred yards into a moss, where it flopped down on three eggs. Carrie also once saw a hen suddenly leave a trip and run down the slope to lay her first egg in a partly-made nest. The other hens stood waiting on a stony ridge some fifty yards away. After laying, the hen returned to the others, and all then flew noisily over the hill. On neither occasion did the territory-holder resist trespass; but brooding cocks usually do attack and eject trespassers throughout the incubation period.

At a nest with two chicks and an addled egg, Seton Gordon remarked that 'the father was angry with a pair of dotterels which were feeding near the nest and he attacked them with ruffled feathers'.

For main meals, some brooding dotterels fly far away to feed and occasionally bathe in wet swampy places. These movements can lead to fights. We watched a cock dotterel fly from a nest to a wet, mossy patch which was also the favourite feeding place of another cock whose nest was also roughly equidistant from it. The second cock also now came off and flew at the trespasser. What a fierce fight that was! All flurries of wings and jumps into the air.

There is no inflexible rule about the location of these feeding grounds, which are sometimes above, at others below, nest-level.

Dotterels only seldom nest beside their main feeding grounds, but their defended nest-territories often provide snacks between main meals. For a few minutes running over rocks and stones, stopping and starting, the hungry brooder catches beetles and spiders on flat or ridge, usually within a hundred yards of the nest. Then they shake or preen themselves and fly or run back to the eggs.

Finnish and Dutch observers had similar experiences. Brooding dotterels regularly fed close to their nests on Värriötunturi and within a radius of seventy-five yards in east Flevoland.

Why do sitting dotterels defend and maintain nesting territories? The cock has to brood his eggs in an exacting climate. Few hens fully share the chores of incubation. The cock thus has to find food quickly while brooding and hatching the eggs. If he quits the nest during hail or sleet storms the eggs may be chilled and the chicks die in them. Drifting snow also quickly buries the eggs. The importance of the nesting territory or defended areas as a source of food probably varies, but some cocks regularly feed close to their nests. In 1953, for example, when dotterels were breeding almost in clusters on the Drumochter tops, I located three nests through finding droppings and then searching the surrounding hillside.

To sum up. By attacking and expelling trespassers, the brooding cock dotterel ensures the preservation of a food cache close to the nest. In severe weather this small extra supply or reserve is likely greatly to help first the parents, then the chicks. Defence of the nesting territory also deters late settlers, thus helping to maintain an exclusive living-space for the newly-hatched young chicks.

Territorial fighting during the fledging period

For dotterel chicks their first forty-eight hours are vital. On the first day the cock broods them in the nest, but he also has to feed himself as well as keep them warm. Then, weak and almost helpless against predators, the chicks totter from the nest, their father shepherding them, leading them to food, and periodically brooding them. They now rely on his knowledge of shelter, food, and hiding places, as well as in their own innate capacity to feed themselves. But their cryptic colour patterns and their father's distraction behaviour are not enough. Only real knowledge of the environment allows survival in a world with foxes on the prowl and crows, jaegers and gulls searching overhead. By knowing all that his territory contains a cock enables his chicks to live through mid-summer snowstorms. Watson (1966) suggests that 'dotterel and ptarmigan chicks often live because their nesting ground contains big rocks and boulders. Large spaces under the boulders are usually completely free of snow and the soil there is not frozen. . . . Midges and other *Diptera* crowd into these places after summer snowstorms.' Territorial defence thus preserves these dumps against competitive exploitation.

I once watched a cock dotterel hide himself from a party of hill climbers. When the climbers appeared on the skyline the dotterel was running backwards and forwards close to his chicks. Quite suddenly he melted beside a stone. For minutes I lost him and thought that he had gone. But after the climbers had passed by he suddenly rose up and started to run again. Yet the climbing party had walked past within twenty-five yards of him. That dotterel certainly knew his ground!

When his chicks are on the move the cock continues to attack strange dotterels, but he is really attacking those coming near his brood rather than defending a territory. Fighting is now confused. A cock sometimes leaves his eggs to eject another cock whose chicks have strayed towards his nest. Or the father of a brood tackles another cock which is on his way back to eggs.

Cocks with chicks frequently attack trips of hens. On 30th June, 1954, a Norwegian observer watched three hens land near a cock whose three small chicks were running nearby. Every time that the hens came within fifty yards of his brood the cock flew at them. Time after time he did this until at last he and the three hens took wing and flew away. A few minutes later he came back and gathered his brood. Soon afterwards, however, a hen flew in and again tried to approach the brood. For several hours she stayed near them, but the cock never allowed her within thirty yards. Judging by the behaviour of other pairs, this hen was possibly the mother but the father would not accept her. As I shall later describe, the hen, although taking little part in rearing small chicks, often tries to join the family party. But, treating her as a trespasser, the cock usually drives her off, thus conserving much needed food for the chicks. Later on, when the young are stronger and more mobile, he accepts the hen into his little group. By this time the social impulse is stronger and the senses of the second adult are now at the service of the chicks.

How important is territory during the fledging phase? Defence of space is probably only meaningful in the few years that dotterels are nesting at really high densities. On the Drumochter Hills in 1953, for example, when there were eleven nests on a single block of hill, six were spaced about two hundred yards apart in an almost straight line along the 2,900–3,200 foot contours. The territorial régime involved during incubation certainly seemed to continue into the fledging period. By standing beside their broods, the different cocks acted as 'distance threats' to others, but the parents fought and bickered when the chicks strayed into their neighbour's living-space.

Earlier territorial behaviour and discipline thus probably assisted parents to obtain fair shares of limited resources. Two hens started to lay after chicks had hatched in almost all other nests. We also recorded non-breeding dotterels of both sexes in full feather. I doubt, however, whether territorial fighting alone caused this non-breeding surplus. Territory is only one of the dotterel's dispersion mechanisms. Status is usually settled at group level. Noisy, volatile group movements also probably determine the general pattern of dispersion. Territory only becomes significant shortly before the pair selects a nest-site and the hen lays the first egg. Thenceforward, territory is valid.

Territorial and replacement clutches

About one hen dotterel out of four replaces a lost clutch. Each hen, however, usually lays her repeat in a different part of the original nest territory. The pair thus perhaps moves far enough to baffle predators, but continues to hold a living-space which it already knows well. In the Cairngorms, eight hens laid their repeat clutches approximately 150–200 (3), 250 (2), 300–350 (2) and about 500 yards away from their first nests – a mean of just under three hundred yards. In the West Grampians, however, a hen laying unusual eggs shifted approximately fifteen hundred yards for her replacement clutch. In this move she crossed the territories of at least two other pairs. I do not know, however, whether she remained mated to the cock which was robbed of her first clutch.

I have no evidence that territorial behaviour has actually limited the numbers of pairs breeding in the Cairngorms. Even in the good years pairs certainly did not nest in every territory occupied in previous years. In 1953, however, the dotterel's dispersion mechanisms probably did restrict the breeding populations on the Drumochter Hills.

Summary

1. Dotterels pair from mating groups, but the newly formed pairs contend for exclusive territories. This territorial system thus helps to

disperse the breeding populations and later reduces abortive fighting during courtship and egg-laying. Two birds can thus woo and adjust and reconcile their sex-rhythms. Within the territory the hen can choose a nest-site and produce her eggs and the cock can also generate his brooding drive.

2. Dispersion reduces the risk of predation. A predator is less likely to discover and destroy well-dispersed nests. An intimate knowledge of hiding places within the territory also helps parents and chicks to escape from predators.

3. Defence of territory during incubation deters late settlers and helps to maintain a food reserve close to the nest. This is important to the brooding bird in storm and blizzard and to chicks which have to feed themselves.

The balance between habitat and population is almost always decisive in determining territorial flexibility. Dotterels and the phalaropes have evolved quite different systems of dispersion, but all promote the survival of unorthodox animals in demanding environments.

CHAPTER 8

Nest and Nest-Dances

Nest-dances

BEFORE she has won a mate, a hen dotterel sometimes begins to make scrape movements. Running quickly over the high field, she suddenly bends forward, sinking to the ground with her creamy-white and cinnamon undertail-coverts fluttering in the wind. Then, almost as if photographed by a camera in slow motion, she starts to make a scoop, lying on her breast and turning partly round in moss or tussock. At first these incomplete movements remind you of a warbler toying with leaf or grass, or a crossbill breaking off and dropping a twig from a pine tree. In this phase the hen's scrape movements are incomplete and uncoordinated. She soon seems to lose concentration and runs away, perhaps feeding nervously or preening herself, pecking at the ground, or giving the appearance of carrying out these actions. Later on, after cock and hen have discovered one another, they twinkle over the ridges, first one, then the other, in no fixed order, starting to make depressions. They now make these movements in scores and in hundreds of different places, but they never seem able to form a real hollow or scratching. A few days later, however, the dotterels' nest-dances have more meaning. Cock or hen now eagerly and energetically rotates, pressing and hollowing with its breast, often scratching away with its feet, and calling up its mate, leaving the nest only to allow its partner to take its place and continue working in the same scoop. Sometimes the cock calls to the hen, at others the hen signals to the cock. The initiative depends, I believe, on which bird has the stronger drive in a particular phase. In the same spasm of nest-dancing, the two dotterels make these movements twenty, or even fifty yards apart, or they meet and work at the same nest. You walk over to inspect their work, but there is nothing to be seen. The dotterels are still unable to complete what they are trying to do.

The next day, perhaps, you notice a subtle difference in their behaviour. They now no longer fly from one patch or ridge to another and they do not casually quit perfectly good-looking nesting ground. For ten minutes, or more, the pair work at the same scoop, often scratching away with their feet as they move round about. Then, with her back to the cock, the hen sometimes tilts forward, breast to ground and tail up, and gradually she leans still farther forward – much as she did when inviting the cock to mount her. Her tail feathers fluttering, she bends down and apparently pecks in the bottom of the scrape. The cock then takes over and replaces her, perhaps jostles her from the nest, precisely copies her actions there, and then shuffles away. Standing just in front of him, the hen tosses or jerks blades of grass or bits of moss over her shoulders towards him. Meanwhile, the bird in the nest leans backwards or sideways, awkwardly working these bits and pieces into place in the nest-cup that it is enlarging. The two birds sometimes stand up and face one another, both raising their wings and fanning out their tails. Over and over again they ripple, or the cock perhaps sings his almost linnet-like song. Nevertheless, although the dotterels sometimes spend an hour or more at one particular scoop, the hen does not always lay in it. Suddenly, and apparently quite casually, she may run away to another scrape – one that you had quite forgotten – and there finally lays her first egg after joining with her mate in the colourful egg-ceremony.

Early in the breeding season sexual rhythms are often slow and unadjusted. The same psychological and physiological difficulties, however, seldom inhibit the building of replacement nests and the laying of repeat clutches. A hen dotterel, for example, sometimes lays the first egg in the repeat clutch on the sixth day after she has lost her first set. The firm pair-bonding no longer exacts or demands protracted mutual or self-exhaustive displays.

Here is a time schedule for a replacement clutch.

14th June: eggs lost
15th June: pair re-courting and displaying on same territory
18th June: full courtship and nest-dances
20th June: hen laid first egg of repeat clutch

In the dotterel's summer world with its short nesting season and brief food-peaks, unnecessary love-making and irrelevant fighting are possibly equally harmful. Failure to adjust could lead to failure to breed. The hen dotterel comes into season ahead of the cock; at first, therefore, she tends to dominate and pursue. Nest-dances assist engaged or courting pairs to adjust and reconcile their respective passions. They reduce tensions, harmonise rhythms, and thus help the pair bond to endure. Only one dotterel couple, in my own experience, separated after serious dating in this peculiar ballroom.

Do the nest-dances excite passion? I believe that they do. In courtship the hen attracts the cock by running in front of him to display her bright undertail-coverts. Standing up in the scrape, or just outside it, she leans forward and then shows off the same sex-exciting pattern. Sitting on the nest, the hen's white-rimmed brown tam-o'-shanter is also

3. Three typical eggs in a characteristic West Cairngorms nest.
Dotterels line their nests with scraps of vegetation often taken from
immediately around the nest scoop, but they also carry in small
bits of grass and lichens from the adjacent hillside.

4. *Above:* A nervous dotterel with flattened posture and withdrawn neck. *Below:* An undisturbed and relaxed brooding dotterel. The broad white eyestripes and pectoral band are clearly shown in this photograph.

possible sexual manna. Nest-dances often end in copulation. Dotterels thus resemble parrot crossbills, which so often mate immediately before or after a stint of nest-building.

Table 2 gives details of nest-dances of some other waders.

Many other waders have equally beautiful nest-dances, with common elements, but different patterns. For each species natural selection has determined the choice.

Material-tossing ritual

I have described how dotterels and other waders bend forward, pick up and then jerk small objects – bits of moss and grasses – over their shoulders. To dotterel, red-necked phalarope and lapwing, this material-tossing is a formal rite in their nest-dance patterns. In other emotional contexts, waders and other birds also throw bits and pieces over their shoulders. During their nest-dances, either dotterel, usually with back to its mate, breast almost to ground, tail almost vertical, bends forward and pecks in the floor of the nest. Then, after the other bird has taken its place, it walks away and jerks bits over its shoulders. There is a still more elaborate interplay, immediately before the hen lays her first egg. One cock dotterel used material-tossing activities as one of his distraction displays. This bird, with one youngster about seventeen days old, chittered and half-turned, breast to ground, and flicked moss over his shoulders.

Dotterels, lapwings and other waders partly line the nest-scoops with bric-a-brac tossed over their shoulders. Ritualised material-tossing during nest-dances also possibly has real survival value. The 'dancers' are highly excited. Material-tossing canalises aggressive drives, inhibiting fighting and helping to maintain the pair-bond. I doubt, however, whether all these patterns have the same origin. The material-tossing actions of dotterels and lapwings, for example, possibly evolved from their bathing behaviour, in which they bend forward and jerk blobs of water over their shoulders.

Choosing the nest-site

I have described the nest-dances of dotterels, but have not discussed whether the hen deliberately chooses a nest-site after testing competing alternatives. Does the hen choose her nest at the last moment? I have only one such record. Casual nest-site selection behaviour is thus apparently abnormal. A pair of dotterels sometimes spends an hour or more in displaying and interchanging at a previously prepared scoop before the hen settles down and lays in it. Nevertheless, although

particular hens sometimes nest for several years in the same territory,
I have never found two clutches in the same scrape in consecutive
years. They thus differ from greenshanks, oystercatchers, golden
plovers and stone curlews, some of which occasionally lay in the same
scoop in more than one year.

Nest-site

In Scotland dotterels usually lay their eggs in small hollows of varying
depth, scratched out or flattened down in patches of woolly-fringe moss,
in tussocks or carpets of deer sedge or mat grass, or in three-pointed
rush among gravel and stones. In the Drumochter Hills I have also
occasionally found nests in prostrate heather. We have found a cock
sitting in a patch of pink moss campion blossom – a beautiful and
unforgettable sight.

In the central Grampians I have found a good many nests quite close
to march or deer fences. In 1953, for example, nine out of eleven nests
on one hill group were within a hundred yards of an old fence. The
nearest was less than ten and the farthest a hundred yards away.

In the central and east Grampians, between 1964 and 1969, Tewnion
also found nine nests near old fences, the nearest about thirty yards,
eight within a hundred yards, and the farthest about two hundred yards
from fence-posts. In the east Grampians Ratcliffe photographed a
dotterel sitting beside a fragment of old fencing, almost like one of
those greenshanks which habitually nest close to snow-fences in north-
east Sutherland. Are these significant adaptations? Tewnion suggests
that dotterels use fence posts as possible landmarks in mist. This
practice is certainly a characteristic of rather featureless fjellmark. In
the Cairngorms I have seen a dotterel, off nest, apparently floundering
in thick mist. We need further research.

Many other dotterels' nests are beside or among stones. A few also
lay their eggs on islands in quite large screefields where the nests are
sometimes placed beside sizeable rocks.

Dotterels sparsely line their nests with mosses, lichens, blaeberry and
crowberry leaves, grasses, small granite chips, and sometimes a few
possibly casual ptarmigan or mountain hare droppings. While brooding,
the cock often stretches forward and picks up or pulls out bits of moss
and blades of grass which he places on the edge of the cup or works into
the lining. Yet a few pairs fail to do this, brooding their eggs in almost
completely unlined nest-cups. The sitting dotterel also sometimes sheds
a few small feathers which soon become part of the lining.

Dotterels seldom move far to collect or pick up nest-stuff, but they
do not always line their nests with material immediately around the

scoop itself. I have watched them run in with grasses or lichen in their beaks and have found items in the lining that were absent from the surroundings of the nest.

Table 3 gives precise descriptions of nests and nest-sites in Britain.

In Norway, Sweden and Finland, dotterels make their scoops among crowberry and reindeer moss, lining them with reindeer moss and a few blaeberry or dwarf birch leaves (Blair, *pers. comm.*).

Close to the Yenesei River in U.S.S.R., Maud Haviland found several nests among marsh grasses in willow scrub. Two nests in almost snipe-like situations were on small hills on the edge of the tundra. One was in a marshy hollow, the second on a hillock rising sixty feet from the swamp 'in soil as dry and stony as the mountain top it resembled in miniature'. In West Taimyr, on the contrary, Walter found two nests in rocky debris in one of which the eggs were lying on pebbles.

Dementiev describes dotterels' nests in U.S.S.R. as small hollows, about nine centimetres in diameter, but deeper than those of grey or golden plovers, with linings of grass and the leaves of Crowberry (*Empetrum*) and other plants.

In Holland one pioneer nesting pair scratched out a scrape in a bare beetroot field and the second laid its eggs in the middle of a row of potato shoots. Five other pairs have nested in flax, two more in sugar beet, and one each in wheat, peas and potatoes.

In Austria Franke found his six nests near the tops of the highest mountains; all were shallow scrapes lined with lichens and situated in low alpine pasture grass. Several nests were sited near or beside embedded stones.

Some dotterels nest beside marks on almost stoneless moss carpets; others equally do not. Is nesting beside a stone an adaptation or is it entirely casual or accidental? The greenshank makes an excellent comparison. With a few exceptions, greenshanks nest beside marks – fragments of dead wood in forest-bogs or close to snowfences, or stones or rocks on treeless moors or flows. If marks have cryptic value to the dotterel, why have selective pressures been so indecisive? Selection apparently does not discriminate against markless hens. Summer snowfalls and blizzards, however, are dangerous hazards. Some cocks continue to brood; others are forced off their nests and then desert their eggs. In exceptionally exposed habitats, therefore, the shelter of a rock or stone is possibly of real advantage. There is certainly a marked microclimatic effect to the leeward of a projecting stone (Ratcliffe, *in litt*). This possibly explains why most nests are between low stones in Finland.

On the Cairngorms and Grampians, dotterels usually nest on high ridges and summit plateaux rather than on slopes and shoulders. In years of exceptionally heavy snowcover, however, a greater proportion of early nesting pairs tend to nest on slopes. This particularly happens in years when sun and wind have exposed or melted larger snow-free patches on hillflanks than on summit ridges. Of seventy-six Cairngorms nests, forty-nine were on flats, ridges or plateaux, nearly twice as many as the twenty-seven on slopes. On the Drumochter Hills, forty out of fifty-three nests were also on summit plateaux, about three times as many as the thirteen nests on slopes. I have insufficient records for England, but although dotterels have nested on slopes as well as on tops, the majority were placed more or less on level ground. Of Ratcliffe's four recent nests, three were on summit flats, the fourth on a gentle summit slope.

In Scandinavia most dotterels likewise probably nest on the tops but there are many exceptions. At Anmarnås in north Sweden, J. Cudworth found a nest with three eggs on quite a steep hill shoulder. This slope, facing east-south-east, also contained some flat areas of sparsely vegetated and eroded soil. The birds probably nested there because the snow had melted earlier than on the higher flats.

CHAPTER 9

Egg

The Egg Ceremony

ON 11th June, 1934, we were watching the dotterels Meeson and Sheila. At exactly five minutes past one o'clock Sheila went to a rather large nest-bowl, in which she stood, slowly picking up little bits of moss and carefully placing them on the rim of the cup, or she peered down and pecked at the bottom of the bowl. Just two minutes, and then Meeson took over from her. He now silently turned round and round, while Sheila stood just in front, tossing bits of moss over her shoulder towards him. For almost an hour Meeson and Sheila almost continuously interchanged. Then Sheila settled down and Meeson turned and faced her less than a foot away. Meeson now twittered to Sheila and she twittered back. Then he ran away, crouching down in the shade of a rock. Sheila first fidgeted and then sat more closely, and once or twice she half-rose and peered down into the nest.

At 2.58 Sheila stood up and then sat down again. Meanwhile, Meeson occasionally called and at 3.40 he ran towards the nest; as he reached it, Sheila flew away and Meeson carefully lowered himself into the scoop. We walked over. Meeson fluttered off. The egg was there.

At 9.23 that night we again looked into the nest. The dotterels were away and the egg cold.

On 9th June, 1940, Carrie and I watched another hen lay her first egg on a stony shoulder below a tableland. The two dotterels were almost side by side in a little hollow between two green ridges. Just above them was a cluster of white quartz stones embedded in the hillside. The time was 16.20. The dotterels occasionally wiggled but were extremely quiet and broody. Suddenly the hen ran to a scrape and started the ploy. After a few minutes the cock gently pushed her off. For twenty minutes he now alternately sat or half-rose, or stood over the nest, throwing bits and pieces over his shoulder in the usual way. At one time the birds gave up and ran uphill, thus giving us a chance to look at the still empty scrape. But soon they were back again. How restless the hen now was! Sitting rather high, she built clumsily, picking up grasses and poking them into the cup with sideways movements. At last she settled down, but once stood up. A few minutes later the cock took her place.

The hen ran a few yards and then, with rather slow and laboured wing beats, she flew uphill and landed beside a stone where we could just see her head against the skyline. The cock still brooded but, when the hen started to call and then flew away, he followed her. We went to the nest. The egg was still warm from the hen's body.

Hundreds of nest-dances thus reach their climax in this egg ceremony, whereby the hen finally chooses and the cock accepts a particular nest-scoop. Dotterels usually only employ this ritual when the hen is laying her first or second egg. But one hen laid her first egg with no ceremony at all, and in the absence of her mate (p. 58).

I have twice seen the third egg of a clutch laid but the birds then behaved differently. On 12th June, 1948, from noon onwards for nearly an hour, a hen pecked at a tussock of grass and alternately shook and then dropped a fox's dropping. Suddenly the cock, which had been calling from a ridge above the nest with its two eggs, ran down and then turned towards the nest. The hen followed him and for a few minutes squatted beside the nest as he sat on it. Then, calling softly, she jostled him off it. After he had flown away she brooded for nearly an hour before half-rising and then sitting down as if she had laid. A few minutes afterwards she called softly and the cock ran up. Immediately the hen flew off in heavy laboured flight and landed on a higher ridge. The cock now went to the nest, turned the eggs, and started to brood. I let him settle, then flushed him. The third egg was laid.

At 10.45 the next morning the hen was again close to the nest; but she did not visit it during the next two hours. I have summarised other observations on egg-laying behaviour in Table 4.

Egg-laying intervals and times

Most hen dotterels lay their eggs at intervals of between twenty-four and thirty hours. Lack of food or severe weather probably slows up the process of egg formation and accounts for gaps of forty-eight hours or more.

Dotterels do not lay their eggs at any particular time of the day. In the Cairngorms one hen laid her third egg at about 07.00 and another her third at about 19.23. At least thirteen out of nineteen eggs were laid in the afternoon. All Scottish data are given in Table 5. In his study area, where there are twenty-four hours of daylight in June, Pulliainen also found that eggs were laid at all times of the day. He records laying at 00.30, about 04.00, 08.20, 11.15 and 20.00. The average interval recorded between eggs was 30.8 ± 2.1 hours (S.D. = 6.6). An Austrian hen laid her third egg between 05.45 and 06.30. After abandoning her first egg, a Finnish hen laid the first of a second trio 44.5 hours later in another scrape.

Laying season in Britain

In the Cairngorms, dotterel, ptarmigan and snow bunting generally have slightly different laying seasons in any one year. In every year for which I have reasonably accurate records, the earliest ptarmigan laid her first egg before the earliest hen dotterel had laid hers. In eight years out of ten the snow bunting was the latest to begin laying. Exceptional snowcover, late thaws, low temperatures or severe snowfalls in May delays nesting and laying. On the other hand, warm sunny weather in the second or third week of May often seems to trigger off laying by waves of hen dotterels, but there is almost always a considerable gap between the first egg of the first clutch and the first egg of the latest clutch.

Exceptionally mild sunny weather in the last week of April and first week of May favours early nesting and laying by ptarmigan. An abnormally warm spell in the third week of May possibly induces snow bunting to start site-selecting and nest-building. Climate, however, is only one factor. Watson (1965) has shown that the ptarmigan's breeding season is probably correlated with the growth of blaeberry (*Vaccinium*) and its other food plants. Changes in weather, however, largely determine growth. Late thaws and May snowfalls are probably less decisive factors. Warm weather and lack of snowcover in the first half of May also probably activate hill insects and spiders, thus assisting some hen dotterels to form and complete their first eggs.

Temperatures alone are hardly likely to account for all variations in the same year. 'The availability of food for the female prior to laying' influences the date of laying in great tit, common swift and cuckoo (Lack, 1966). The great tits apparently need four and the swift five days to form and complete an egg. Dotterels, therefore, presumably respond to insects and spiders activated by a few warm days in May and ptarmigan to the growth of their food plants. If a hen dotterel misses the early tide she is possibly unable to complete her first egg. This helps to explain why dotterels lay in waves. Presumably, therefore, a hen which has already laid an egg must attempt to complete its clutch, even if hard weather intervenes.

Table 6 gives the dates in the years when I have reasonably precise records of the first egg laid in early clutches. The earliest first egg record in the Cairngorms varied from 9th May in 1940 to 13th June in 1932 – a mean date of 27th May (thirty-six years). In almost every year there was a considerable gap between earliest and latest hens. Disregarding replacement clutches and second clutches of the few poly-androus hens, I find that the average date when the last hen laid her first egg in the Cairngorms was 10th June (seventeen years). Even in exceptionally early years, such as in 1933 and 1940, there were long intervals between the first egg of the earliest and latest hens. In the warm May of 1933, for example, Blackie, the earliest hen, laid her first egg about 12th May and the latest hen laid her first on 3rd June. In the notoriously hot Hitler's Summer of 1940, we found the nests of seven pairs of dotterels; six of these hens laid their first egg on or about 9th, 10th, 13th and 18th May and 9th and 15th June. I have records for twenty-one years in which the intervals between the earliest and latest hen varied from under a week in a few years to thirty-seven days in 1940 – an average of fourteen days for the twenty-one years. The records for all years, however, are not equally extensive.

In the central Grampians the average date when the earliest hen laid her first egg is 22nd May, five days earlier than in the Cairngorms. In the eleven years for which I have records, 2nd June was the average date for the first egg of the last clutch. This gives an average gap of eleven days between the earliest and latest hens.

The earliest central Grampian hen laid her first egg on about 14th May, 1911. In 1953 the first hen laid her first egg on about 18th May and the last on 11th June. As in the Cairngorms, I have disregarded replacement clutches and the second clutches of possibly polyandrous hens.

Records for the English fells – Cumberland, Westmorland and Northumberland – are few; but I have been able to ascertain the *approximate* date on which the earliest egg was laid in eighteen different years. The mean date was 23rd May. It is perhaps surprising that the laying season in the English hills is not significantly earlier than in the Grampians, which are so many miles farther north, but records in England are comparatively few. Dotterels nesting in England, however, probably often lack social stimulation. The somewhat later laying season of the Cairngorms birds compared with the Grampians and the English Lakeland nesting groups is interesting. In some years the additional 500 to 1,000 feet in altitude is possibly important in relation to snowcover and lateness of thaw. In the early summer of 1970, however, the first egg was recorded in Lakeland about 14th May, in the Cairngorms on 13th May, and in the east Grampians about 24th May. This year produced an exceptionally rapid thaw.

In early years height apparently has little effect on the dotterels' laying season. The hens which laid in the first half of May in 1911, 1933 and 1940, all produced their clutches in the central Cairngorms on slopes or ridges over 3,600 feet above sea level. In 1933 and 1940 the many early broods of ptarmigan on these same tops were among the earliest recorded. In 1951 most dotterels on the Drumochter Hills, nesting on, or slightly above, the 3,000 foot contour, laid at usual dates. In the central Cairngorms, on the contrary, half a dozen pairs had not nested by 21st June. This was, however, a year of exceptional snow-cover in the central Cairngorms, particularly above 3,500 feet, where most of the dotterels usually breed. In 1955 dotterels in the central Cairngorms, nesting on or below the 3,000 foot contour, started to sit a little later than average. 1968 was another exceptionally late year, with heavy solid snowcover on the high tops of the Cairngorms. On 15th June Sandy Tewnion found a cock brooding on a nest with one egg in which two hens ultimately laid. On the same hill, on 26th June, Adam Watson, Sen., and I watched another pair running about in the manner of birds that were just about to lay, and on 11th August Adam Watson probably found the brood from this unfound nest, nearby.

Table 7 gives notes of the laying season of dotterel and ptarmigan in early and late years. In many of these years we found that ptarmigan nesting on slopes and screefields in the Lairig Ghru nested earlier than those on the high tops and ridges. Watson (1965) also found that 'plants at 1,100 metres were usually 10–14 days later in growing (more in snowy years) than at 760 metres and there was a similar difference in the breeding season of ptarmigan at these altitudes'. In the Cairn-gorms, however, height is a less decisive factor in the dotterel's than in the ptarmigan's laying season. Heavy snowcover, however, does delay nesting and sometimes forces pairs on to lower slopes and ridges.

Laying season outside Britain

In Norway dotterels vary. In southern Norway some hens are not early nesters. In 1953, for example, hens on Bremangeroy, an outer island off Nore Fjord (lat. 61° 50′) laid her first egg in the first week of June. In 1955 another hen on the island of Averöya off Nordmore was equally early. 1955 was thus possibly an early breeding year. At Nore Fjell in Buskerud, a hen laid her first egg on about 14th May – an exceptionally early date for a Norwegian dotterel. Hens in Nordmore clearly have as protracted nesting seasons as those in Scotland. A second hen in 1955 laid her first egg on 16th June.

Konservator Edvard Barth has found fifteen dotterels' nests on the Dovre Fjell, all with three eggs. In 1946 three clutches hatched between 23rd and 25th June. In all three nests, therefore, the hens possibly laid their earliest eggs in the last week of May. A 1951 clutch was later; the chicks hatched on 3rd July, so the first egg was probably laid round about 4th June. Yet the earliest of the seven nests (the eggs hatching on 23rd June, 1946) was laid on a fjell about 4,900 feet above sea-level. The latest nest (chicks hatched on 3rd July, 1959) was at 4,000 feet – roughly 900 feet lower down.

In East Finnmark, T. Schaanning found that hens seldom began to lay until the third week in June, but he recorded a brood of newly hatched chicks on 2nd July, 1913. That hen must have laid her first egg very early in June. Professor W. Hobson also tells me that on a hill above the Varanger Fjord three clutches were completed respectively on or about 3rd, 8th and 13th June. This means that the earliest egg was laid on or about 31st May or 1st June – an early date for a dotterel nesting north of the Arctic Circle.

In Sweden the earlier hens begin to lay in the second week of June. Ulf Houmann kindly gave me notes on clutches taken in twelve different years. The earliest clutch was about ten days incubated on 13th June, 1960. This hen thus probably laid her first egg about 31st May. This was remarkably early as Houmann tells me that there was heavy snow-cover on this particular hill in that year. He suggests that 12th June is the average date for a full clutch in Sweden.

Some Swedish dotterels nest much later. Hans Rittinghaus records a first egg laid on 2nd July, 1959, and P. O. Swanberg (*in litt*) watched a dotterel which laid her first egg about 11th July.

In Finland Professor Pontus Palmgren and Torsten Stjernberg have given me records of over thirty clutches, which vary from three fresh eggs taken on 29th May, 1934, at Ektoniô to a late clutch of three slightly incubated eggs collected on 30th June, 1909, at Enari. The last hen had probably laid her first egg in the fourth week of June.

Dr O. Hildén (*pers. comm.*) has given me records of ten nests in Utsjoki, North Finland. The earliest egg was laid about 5th June, 1967, and the first egg in the latest clutch about 13th June, 1965. In 1969 Pulliainen recorded the earliest egg on 6th June. Six pairs hatched their clutches between 5th–8th July and the seventh between 11th–12th July – a spread of about six–seven days. In Finland, therefore, there is also a considerable annual and possibly individual spread in egg-laying.

In 1867 and 1904 dotterels began to lay about the end of May in the Carpathians. In 1960 and 1961, however, they apparently nested a little later.

Between 25th and 27th June on the Kurchum Mountains, U.S.S.R., Polyakov recorded almost fresh clutches, clutches on the point of hatching, and small downy chicks. The dotterels had thus started to lay in waves from late May onwards.

On 20th July between Chulyshman plateau and the River Dzhity-dei

Sushkin found well-feathered young dotterels, some already able to flutter. In these locations, therefore, the earliest hens must have started to lay in early June.

In the upper reaches of the River Chagan-burgazy in Sailyugem, a young chick about two to three days old was recorded on 6th August. This hen must have clutched towards the end of the first week in July. In Taimyr, Walter apparently found four highly incubated and three slightly set eggs on 11th July. These clutches were thus possibly complete in the third week in June and the first week in July respectively (G. P. Dementiev, 1954).

We have a few records on the laying season of dotterels on the Continent of Europe. On 30th June, 1946, Josef Maran found a dotterel with three two-day-old chicks in the Riesengebirge, Czechoslovakia. The first egg was thus probably laid about 30th May.

In Austria the earliest hen probably laid her first egg about 29th May; the second on 1st June; and the third on 12th June. The fourth nest probably belonged to a polyandrous triangle. In 1952 one hen also laid her first egg on 1st June. The laying season in the Seethal Alps is thus roughly the same as that in the Cairngorms.

On 19th July, 1952, a chick in the Italian Abruzzi had probably hatched from an egg laid in the third week of June.

Between 1961–68 the pioneering pairs which nested in Holland were remarkably early. The first eggs in seven clutches were laid between 5th and 11th May. These dotterels thus quickly adapted their laying season to low ground habitats.

Egg

Dotterels often lay beautiful and attractive eggs, in shape ovate or pointed ovate, typically of common tern *S. hirundo* shape, occasionally ringed plover *C. hiaticula* shape. Waders laying clutches of three or less tend to produce eggs of more ovate than pyriform shape (*e.g.* dotterel, Kentish plover, oystercatcher and stone curlew). Is this largely an adaptation to facilitate brooding? The position of eggs in the nest is possibly important, but waders sometimes successfully brood and hatch double-clutches (D.N.-T., 1951; 115, 118). Woodcock and avocet also usually lay four eggs – broad or blunt ovate in woodcock and blunt ovate or subpyriform in avocet, without known survival disadvantage. Further investigation is needed. The ground colour of dotterels' eggs varies from olive and umber brown to clay or stone colour. A few eggs are genuinely green or greenish: many fresh eggs have an olive or greenish cast which soon fades. Blackie and a few other hens also laid eggs with a deep reddish or tawny-brown ground colour. I have also found other clutches with single eggs of pale blue or greenish-blue colour, but never an entire clutch. Nevertheless, B. Nelson, M. Everett and A. Tewnion have all found in Scotland complete clutches of blue or greenish-blue eggs. In 1971 G. Sutherland also found a clutch of immaculate blue eggs in the East Grampians.

Markings consist of black, brownish-black, and red-brown blotches, caps, zones, streaks and squiggles. Undermarkings vary from violet to light ash-grey.

Jourdain gave measurements for 100 British eggs:
Average: 41.10 × 28.87 mm.
Maxima: 46.7 × 31.3 and 44.5 × 31.5 mm.
Minima: 37.15 × 29.0 and 41.6 × 27.4 mm.
The second hen in Tewnion's 1968 trio laid an even smaller egg – 37.0 × 26.0 mm.

Tables 8 and 9 summarise data on clutch-size in Great Britain, Fenno-Scandia and Europe.

1. Average clutch-size in England (2.58) is less than in Scotland (2.91). In England, dotterels have laid clutches of less than three eggs in 36.4% of recorded nests as against 8.6% of those found in Scotland. This suggests that dotterels in England, *particularly in the first half of this century*, were possibly breeding from a less fertile strain or were then occupying fringe or marginal habitats. It is noteworthy, however, that since 1956, during the dotterel's revival in England, all eight nests found have contained three eggs. Compare Tables 10 and 11. Collectors possibly occasionally took incomplete clutches, but this seldom happened.

2. Average clutch-size in Fenno-Scandia (2.98) is not only slightly higher than that for a larger sample in Scotland (2.91) but there were four clutches of four eggs (2.5%) in Fenno-Scandia as against the solitary Scottish 'four' (0.2%). Again, 8.6% of dotterels in Scotland laid clutches of one or two eggs as against 3.9% in Fenno-Scandia. The combined Anglo-Scottish proportion (12%) of two or one is even more suggestive.

Dr Bob Moss, of the Nature Conservancy, kindly made a statistical analysis of these data. 'I have examined the Tables of clutch size in dotterel from a statistical point of view. I have done x^2 tests to see what is the likelihood of some of the obvious differences between areas occurring by chance. Obviously there are more 4's and fewer 2's from Scandinavia than from Scotland. The odds against this happening by chance are longer than 40 : 1 ($P<0.025$). The difference between England and Scotland is even more marked, England producing more

"1's" and "2's" than Scotland. The odds here are longer than 1000 : 1 (P<0.001).

'I also tested the difference between the Cairngorms and a hill in the East Grampians, to see if the lack of "1's" and "2's" from the east Grampians was significant. The odds here were between 4 : 1 and 10 : 1 against this being due to chance (0.25<P<0.1). This would not be regarded as significant, although it is quite suggestive. More data are required to be certain one way or the other.'

Tables 8 and 9 clearly show that in different parts of Scotland, and in other countries, the commonest clutch-size by far is three eggs. I know of only six well authenticated 'fours'.

1. 1/4, 22nd June, 1880, Norway. This clutch in Ulf Houmann's Collection is extremely even. Houmann himself has no doubt that this clutch is authentic.

2. 1/4, 22nd June, 1881, Enontekiö, Finland. This clutch is in the Wasenius Collection in the Museum of Helsinki, Finland.

3. 1/4, 4th July, 1909, Enontekiö, Finland (T. Stjernberg, *in litt*).

4. 1/4, 23rd June, 1910, Enontekiö, Finland. Eggs slightly incubated (E. Merikallio).

5. 1/4, 11th July, 1881, West Taimyr, U.S.S.R. (G. Walter). This clutch was probably authentic, but Pleske's data are confusing.

6. 1/4, 10th June, 1955, Grampians, Scotland. (A. Gilpin and J. Armitage). Gilpin (*in litt*) tells me that this was an even clutch and that all four eggs appeared to have been laid by the same hen. This is the only acceptable 'four' recorded in Britain.

On 29th July, 1968, however, Sandy Tewnion (1969) found a nest with four eggs in the Cairngorms. 'I measured the eggs, made a written description and photographed them. Even to the unaided eye one egg was obviously different from the other three. It was smaller, measuring 38 × 26 mm. (smaller than the minimum quoted in *The Handbook*), but a typical dotterel egg, being of a dull olive-green ground colour, spotted with dull brown spots and blotches. The other three measured 42 × 37 mm., 41.5 × 28 mm. and 43 × 27 mm. They were pale blue in ground colour and were more heavily spotted and blotched with brown and black. They had obviously been laid by the same hen and there could be no doubt that here was a standard clutch of three with a fourth egg laid in the same nest by a different hen.'

In Scotland, as well as in England, a few hens annually lay two eggs only. In the Grampians a particular hen laid two eggs in the same territory in 1952 and 1953, but in 1954 she shifted about half-a-mile, but again laid only two eggs. Her eggs were of a peculiar shape and of the same general colour pattern in all three years. Only one out of 402 Scottish nests had a clutch of one egg, but four out of fifty-five English nests contained one egg.

Replacement clutches

Dotterels usually have a short breeding season in a severe environment. Laying, brooding and chick-tending take about eight weeks. This perhaps explains why so few hens produce replacement clutches. Only eight out of thirty-four (23.2%) of robbed hens followed up laid repeat clutches. A hen dotterel, however, sometimes lays the first egg of the replacement clutch on the fifth or sixth day after the loss of the first. Repeat clutches occur in late and early nesting seasons, but are less likely if the eggs are robbed late in the season or are heavily incubated. In U.S.S.R. Dementiev (1954) suggests that, 'in case of loss', a hen sometimes repeats in about a fortnight. Table 12 gives records on the timing of some replacement clutches.

TIMING OF REPRODUCTION IN DOTTEREL AND PTARMIGAN

Adam Watson and Raymond Parr

INTRODUCTION

At the outset we would like to say we think Desmond Nethersole-Thompson has collected a unique set of data. One very rarely gets the opportunity of examining thirty years of data on even a common species, let alone a rare one like the dotterel. We are very glad to have had this opportunity. What we have done in this report has been no more than a very brief and hurried exploration of some of the questions that subjectively interested us among the many possible problems suggested by this set of data. The main questions we asked were:

1. Is the timing of egg laying in first and last clutches of dotterel correlated with early summer temperature?

2. Is the duration or spread of egg laying between first and last clutches correlated with early summer temperature?

3. Do ptarmigan and dotterel show a similar timing of egg laying?

4. Do dotterel populations on different areas lay their eggs at the same time?

5. Is the number of young dotterel reared per old bird correlated with July temperatures and with features of egg laying such as early or late clutches?

THE DATA

The entire collection of data, gathered by D.N.-T. from his own observations and from egg collectors and naturalists before 1933, is described in Chapter 9 and shown in Table 6. In some cases it amounts to twenty-seven and thirty-six years' data from different areas. We have not used all these data for statistical analysis and have concentrated mainly on D.N.-T.'s own observations. D.N.-T. in fact gave us twenty-nine years of his own data on clutches of dotterel, but in Table 13 and in the main statistical analysis we used only the twenty-four years from 1933 to 1963, where he also had similar data on ptarmigan.

The validity of the data has next to be considered. In fact one seldom finds nests of ptarmigan or dotterel with only the first egg, and in most cases the dates are calculated by working back from dates of clutch completion or hatching. This is valid as the data are comparable between years; in fact either the date of clutch completion or of hatching of the first nest could have been used as well as the date of the first egg. The effort of nest searching was similar in each year and the main work was done in the same central part of the Cairngorms. Furthermore, the data in Table 13 are all taken from a similar altitudinal zone within this central area, above 3,500 feet, where both dotterel and ptarmigan breed. Some ptarmigan – but not dotterel – breed down to 2,500 feet, where nesting is much earlier, and this is why we have not included in Table 13 any data on ptarmigan nests below 3,500 feet.

In most years there were also data for the first egg in the last clutch of dotterel found in the Cairngorms. Subtracting the date of laying of the first egg in the first clutch from that in the last clutch also gives us the spread of reproduction for each year. Then there were similar data for first egg dates from the Grampians and from English Lakeland (Table 14), based largely on other people's observations, and we have compared these with the Cairngorms data. Finally, D.N.-T. had observations from various years on the breeding success of dotterel based on the number of full grown young reared per old bird (Table 15). We have compared these data with summer weather, using only those years when more than five adults were seen, to reduce the great variability that occurs with very small samples.

METHODS OF ANALYSIS

The main method we have used is to compare the various sets of data on graphs, look for patterns and possible relationships, and test relationships by correlation analysis. Strictly speaking, some of these tests would be more appropriate for regression analysis than for correlation analysis. However, the main method of calculation is the same in both and in no case were we concerned with using a regression equation to try and predict what future result might occur with the dependent variable on the y axis (e.g. timing of breeding) given a particular value for the independent variable on the x axis (e.g. spring weather). When statistical details are given, the symbol r is the 'correlation coefficient'. A positive value (+) indicates that both variables tend to be high together; an inverse or negative value (−) indicates that the two are inversely correlated so that one is high when the other is low. r^2 indicates the percentage of variability which can be accounted for by the relationship being tested, e.g. two r values of 0.3 and 0.6 give r^2 indicating that respectively 9% and 36% of the variability in these two cases is accounted for by the relationship. P is a symbol for probability. $P<0.10$ means that the probability of the result being due to chance is less than one in ten, <0.05 less than one in twenty, <0.01 less than one in 100, <0.001 less than one in 1,000. 'Not significant' means that the probability of the result or relationship being due to chance is *more* than one in 20. In some cases the figure for r is high but the P figure indicates no statistically significant result, even though in other cases the r value is lower and yet the relationship is highly significant. High r values with P values not significant, happen mainly because the number of samples being tested is small.

RESULTS

D.N.-T. had already thought that dotterel and ptarmigan nested earlier in mild snow-free springs, and one aim of our analysis was to test the hypothesis. We compared nesting dates with the mean temperature in May each year at Braemar, which is the nearest place where weather observations have been recorded throughout the period covered by the data in D.N.-T.'s book.

The analysis shows that the first egg dates of ptarmigan are inversely correlated with the mean air temperature in May at Braemar (r = −0.43, $P<0.05$). This is not a highly significant ($P<0.01$) relationship and Fig. 1 shows a wide scatter of points. The ptarmigan nests so early – in some years the first eggs are at the beginning of May – that one would not expect a very good correlation with the mean temperature

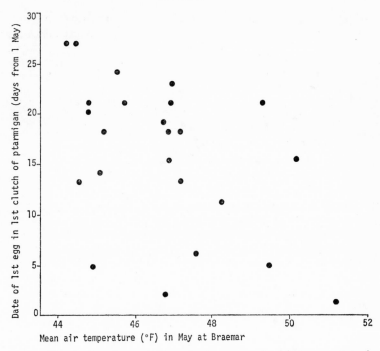

FIG. 1. Egg laying dates of Ptarmigan compared with temperature in May.

for the whole month of May. Other features of climate, such as the snowcover remaining from the winter, or the mean temperature in late April, might also be important. The first egg dates of dotterel are related – inversely (Fig. 2) – to the mean air temperature in May at Braemar ($r = -0.48$, $P < 0.05$). We can conclude therefore that both ptarmigan and dotterel lay earlier in years when May is mild and later when it is cold. This being so, it is not surprising that the first egg dates in first clutches of ptarmigan are correlated (Fig. 3) with those in dotterel ($r = +0.684$, $P < 0.01$). Fig. 3 shows that dotterel are invariably later than ptarmigan, on average about ten to fourteen days later. So, although dotterel are early in years when ptarmigan are early, and late when ptarmigan are late, in any one year they tend to be ten to fourteen days later than ptarmigan. Fig. 2 also shows no evidence of a relationship between the date of laying of the first egg in the first clutch of snow buntings in the Cairngorms with the date for dotterel. However, in this case there were data from only five years for the snow bunting.

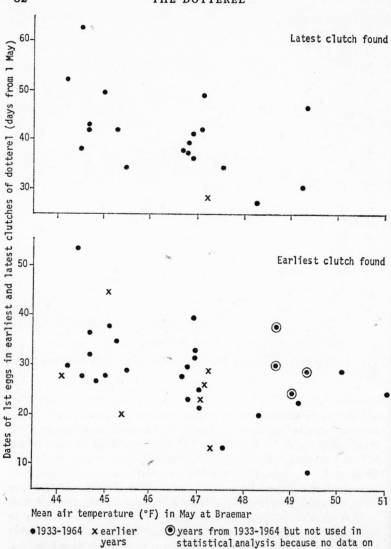

FIG. 2. Egg laying dates for first and last clutches of Dotterel in different years compared with temperature in May.

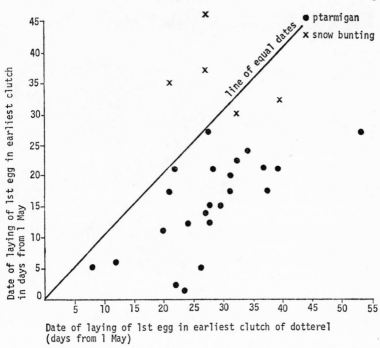

FIG. 3. Egg laying dates for earliest clutches of Ptarmigan and Snow Bunting compared with Dotterel in the same year.

The date of laying of the first egg in the *last* clutch of dotterel is also associated inversely (Fig. 2) with the mean temperature in May, but the relationship is not statistically significant ($r = -0.41$, $P < 0.1$). Nevertheless a tendency is indicated in that result, for laying of the last clutches to be later in years of low May temperatures.

The graphs of these data in Fig. 2 show that observations from earlier years before 1933, gathered by D.N.-T. from egg collectors, fit the general pattern of his own observations from 1933 to 1963. These earlier data were not used in the statistical analysis, however, as we felt they were not exactly comparable (e.g. the amount of effort in searching and the seasonal timing of this effort, may have been different from those used by D.N.-T. himself).

The spread of laying in dotterel (number of days from the date of the first egg in the first clutch to the first egg in the last clutch) is inversely correlated with the date of laying of the first egg in the first clutch ($r = -0.594$, $P < 0.01$). In other words, the later the laying of the first

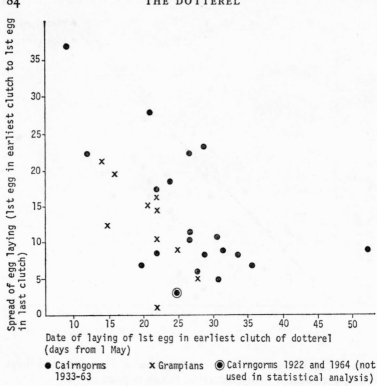

FIG. 4. Spread in dates of Dotterel egg laying compared with dates of laying first eggs in the same year.

clutch, the less is the spread of laying between first and last clutches (Fig. 4). Fig. 4 also shows that D.N.-T.'s observations from the Grampians fit the same general pattern as the Cairngorms data, indicating an inverse relationship between the spread and the timing of the first egg in the first clutch. Possibly there is a tendency for the spread to be slightly less than in the Cairngorms, and for the first egg to be laid slightly earlier, but we did not test this by statistical analysis.

Fig. 5 shows no evidence of any relationship between the timing of the first egg in the first clutch of dotterel in the Cairngorms and timing of the same event in the Grampians. This seems remarkable, as D.N.-T.'s Grampian area is only twenty miles from the Cairngorms. However, dotterel on this Grampian area lived at an altitude of 2,800–3,300 feet whereas those on D.N.-T.'s Cairngorms area lived at over 3,500 feet. The Grampian area was always largely snow-free by mid-May and very

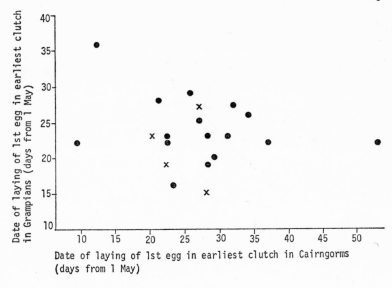

FIG. 5. Dates of first Dotterel eggs in the Grampians compared with the Cairngorms in the same year.

rarely had summer snowfalls that lay for more than a few hours. By contrast, the Cairngorms area was still largely buried by winter snow in mid-May every year and even in late May and mid-June in some years, and summer snowfalls lasting between one and two-three days occurred frequently. The idea that climatic differences may be important in explaining the lack of agreement between timing of egg-laying in the Grampians and the Cairngorms is confirmed by the fact that first egg dates in first clutches in the Grampians are very strongly correlated (Fig. 6) with those in Lakeland ($r = +0.844$, P<0.01). The Lakeland area was at a similar altitude to the Grampians area and, unlike the Cairngorms area but like the Grampians one, was always largely clear of snow by mid-May. In fact, the r value here is very near the figure of $r = 0.847$, where the P level reaches the <0.001 probability. In this case, all data were combined using D.N.-T.'s observations and those of other people. An interesting feature shown by the graph in Fig. 6 is that, within any one year, Grampian dotterel tended to lay several days earlier than Lakeland ones. Yet Lakeland is nearly two hundred miles south of the Grampians and receives a warmer winter, spring and summer. Possibly this suggests a little more evidence to that

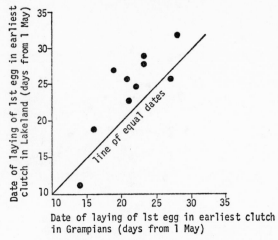

FIG. 6. Dates of first Dotterel eggs in the Lakeland
compared with the Grampians in the same year.

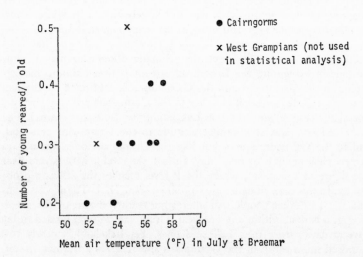

FIG. 7. Breeding success of Dotterel compared with July temperature.

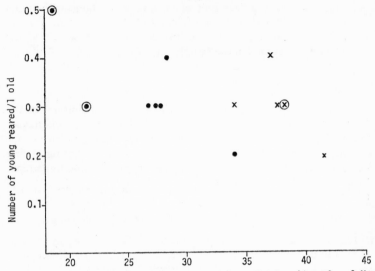

Dates of laying of 1st eggs in earliest and last clutches (days from 1 May)

● 1st egg in earliest clutch, Cairngorms

◉⊗ As above, for West Grampians (not used in statistical analysis)

x 1st egg in last clutch, Cairngorms

IG. 8. Breeding success of Dotterel compared with egg laying dates for earliest and last clutches in the same year.

given by D.N.-T. elsewhere, that the Lakeland population was a marginal one, laying smaller as well as later clutches.

There is a tendency (Fig. 7) for breeding success (number of full grown young reared per old bird or ratio of young to old) to be higher in years when the mean air temperature in July at Braemar is warm. The relationship is not statistically significant ($r = +0.586$, $P < 0.1$), although it superficially looks strong in Fig. 7, but there were only eight years of data and a bigger sample of years will be needed to test this hypothesis properly. The idea behind it was that the young might survive better in warm summers, presumably because they would need less brooding, could thus spend more time feeding, and might have more invertebrate food readily available.

The ratio of young to old shows a suggestion (Fig. 8) of an inverse relationship with the date of laying of the first egg in the first clutch. However, it is not statistically significant although r is -0.635. Again the number of years of data in the sample is too small to make a proper test. Fig. 8 shows a similar hint of an inverse relationship between young to old ratio and the date of laying of the first egg in the last clutch. We did not test this by statistical analysis, however, as the number of years of data in the sample was even less.

SUMMARY

1. The timing of the first egg in the first clutch of dotterel in the Cairngorms was correlated with timing of first eggs in ptarmigan. Within any one year the dotterel laid on average ten to fourteen days later than ptarmigan. There was no evidence that the timing of first eggs in snow buntings was correlated with timing of laying in dotterel.

2. The timing of the first egg in the first clutch of dotterel was inversely correlated with the mean temperature in May at Braemar, which was the nearest weather station. So also was the timing in ptarmigan.

3. The duration or spread of laying between the first egg in the first clutch and the first egg in the last clutch of dotterel in the Cairngorms was inversely correlated with the date of the first egg in the first clutch; i.e. the earlier the dotterel laid, the greater was the spread of laying up to the last clutch.

4. Laying dates of dotterel in the Grampians were strongly correlated with those in English Lakeland, but showed no relationship with those in the high Cairngorms.

5. Breeding success (number of full grown young reared per old dotterel) tended to be higher in years when July was warmer, and in years when first egg dates were early. These two relationships were not statistically significant, but in these cases there were too few years of data to test these ideas properly.

CHAPTER 10

Dotterels at the Nest

NEST-DANCES and egg-ceremonies probably stimulate the cock's brooding drive, but he seldom continuously sits on the first egg. After the hen has laid it, she usually flies away and the cock often accompanies her. Sometimes, however, he later returns to brood the egg, but seldom really warms it, although when flushed he may distract quite strongly. I have flushed two dotterels from one egg in the early morning, but four others did not brood single eggs at night. Carrie once found an egg split by a sharp frost.

Seton Gordon watched a hen feeding close to a nest with one apparently newly laid egg which she was entirely ignoring. In a sudden flurry of hail the cock appeared, flying fast and low across the hill, and immediately settled down to cover and protect the egg. But he was too late. A hailstone had already dented it.

A hen which Gordon watched on 2nd June, 1939 behaved abnormally. In the afternoon she was sitting on her first egg. The next day, at 11 a.m., she was still brooding, but allowed Gordon almost to touch her, merely 'holding up one wing to protest'. On the afternoon of 4th June she was still sitting but had now laid her second egg. 'She allowed my wife to touch the tip of her bill and held up both wings. The heat was clearly distressing her, so I picked a snowball from a snowfield and placed it beside her. She stopped panting and dozed with half-closed eyes. In Glasgow that day the thermometer registered 80° in the shade.'

On 1st July, 1953, I watched a pair of dotterel, with one egg in the nest, change over every few minutes. The hen frequently hustled the cock from the egg and brooded it herself. All the time the two birds continuously twittered to one another.

On 15th July P. O. Swanberg flushed a dotterel (of unknown sex) from one egg, warm, but certainly not at incubation temperature. Some hours later the egg was cool and dry and both dotterels absent. At 14.30 Swanberg left his hide but at 18.30 found the cock was sitting on two eggs. Between 7.30 and 8.55 on the 16th the cock continued to sit on the two eggs, but the hen was still near the nest.

At 14.03 on 2nd July, 1959, Rittinghaus found a cock covering one cold egg. At 14.00 on the 3rd the cock was still brooding one warm egg. At 16.10 he trapped and ringed both dotterels and at 17.45 found that

the hen had laid her second egg. Both dotterels, however, had gone and
the eggs were cold.

A cock often starts steady brooding soon after the hen has laid her
second egg. But two waited until the hen had laid out. On the day
before a hen laid her third egg the two dotterels ran in circles over the
flat for about three-quarters of an hour. Then the hen went to the nest
and started sitting rather fitfully, preening herself and putting odds and
ends on the rim of the scoop. A little later the cock jostled her off the
two eggs and started steady and continuous brooding. I have also
watched a brooding hen whistling softly as she ran up to a cock which
was steadily brooding her first two eggs.

A few other boreal and tundra waders start steady brooding before
the clutch is complete. These include semi-palmated, golden and grey
plovers, and ruddy turnstone, all species in which both partners help to
rear the young; and white-rumped sandpiper in which the hen alone
takes charge of the brood.

Does the dotterel gain from this adaptation? Let us consider the
options. The cock alone, or often almost alone, broods and hatches the
eggs and later tends and rears the chicks. The hen is thus not available
to lead away the brood or tend a precocious chick which has left its
father's warmth. Why then do dotterels start brooding before the hen
has laid out? Although the eggs are fairly frost resistant they are rather
fragile and thin-shelled. Sharp frosts sometimes split and hailstones
break them. That being so, why does not incubation always begin with
the first egg? If the cock started steady brooding on the first egg the
chicks might hatch at intervals of about twenty-four hours, the last
emerging over forty-eight hours after the first. Lack of a second parent
might then endanger the brood. The oldest chick would tend to stray,
perhaps inducing the cock to follow, and thus put the younger chicks at
risk. On the other hand, by sticking to the nest until the last chick is
strong enough, he might lose control of the two rovers.

Incubation starting shortly after the last egg – the normal practice of
most waders – would probably ensure more synchronous hatching, but
would endanger the two unbrooded eggs in severe weather. Steady
brooding beginning after the hen has laid her second egg is thus
probably the best option. All three eggs, moreover, often hatch within
6 to 12 hours, possibly because the cock does not immediately generate
his full brooding-drive. This short hatching period, therefore, enables
the cock to hatch and brood all three chicks and lead them away
together.

Share of sexes in brooding

In Scotland a few hens brood eggs, but we have few precise records. We now have seen 114 nests, but only at seven were we able to *prove* that the hen had brooded. Here are our records.

At 10.26 on 23rd June, 1934, while searching for Blackie's second nest, a dotterel, calling rhythmically, came winnow-gliding over the hill-shoulder. Like a flash, a second dotterel sprang from a higher ridge and flew up towards the incomer. The incoming bird then flashed down to the ridge from which the second dotterel had come. We now searched this ridge without success.

At 17.30 on the 25th we again saw a dotterel fly in *peeping*, and again a second bird rose and flew up towards it. The incomer then swished down to the slope above. We now knew roughly where to search and after exactly fourteen minutes Carrie flushed a large plump hen from the three beautiful eggs which only Blackie could have laid. We now went away, allowing Blackie to settle down, but two hours later she was still brooding. Now, at last, we knew that a hen dotterel was sharing in brooding duties, possibly at night.

We continued to watch Blackie's nest. Between 14.00 and 15.00 on 1st July the cock, a faded little fellow – small, thin, and quite unlike the brilliant and plump Blackie – was sitting. At 20.19 on 4th July, however, Blackie was brooding again. How differently she behaved and how large and bright she looked by comparison with her faded little mate! But we never again flushed her from her eggs. At 12.30 and 20.00 on 5th July, at 17.30 on 7th July and at 13.30 on 11th July the cock was on the nest. Never again did we see Blackie anywhere near her nest. But she had already made history.

On 25th June, 1934, I also found a cock sitting on another nest about 300 yards away. This cock was also a thin spidery little fellow. At 12.38 on 28th June I saw one dotterel run up to this nest and another run away. I then found that the hen was brooding. She chittered, fanned her tail, fluffed herself and beat her wings. I watched her run back to the nest. At 13.15 on 1st July I probably disturbed an exchange when I saw the two dotterels at the nest.

We recorded no more brooding hens until 31st May, 1937. I then watched two hens go on to nests with slightly incubated eggs. Both were large, plump, bright birds, with pure deep-brown crowns.

At 06.47 on 23rd June, 1937, the hen we were watching suddenly stopped and a second dotterel rose almost at her feet. Our bird now sat down. A few minutes later we flushed her from three eggs to which she returned inside ten minutes. At 07.21 the cock returned and landed

about fifty yards above the nest and ran back to it. A few soft whistles
and then the hen flew away.

I have described the strange hen watched in 1941. On 5th to 9th June
she invariably ran rather drowsily from her two eggs, half-heartedly
drooping her tail.

From 1942 to 1948 we never again proved that a hen was brooding –
except, of course, while she was actually laying. At 20.10 on 9th June,
1949, however, I did flush a hen from three slightly incubated eggs.
This hen was possibly a night-brooder. On 2nd June, 1950, a hen was
also sitting on two eggs between 20.50 and 22.00, but on the afternoons
of 3rd, 4th and 5th June the cock was brooding them.

A few other watchers have recorded hen dotterels brooding in
Scottish haunts. From 13.30 to 18.30 on 11th June, 1950, in cold and
windy weather, Harold Auger sat in a hide at a dotterel's nest. At least
fifty yards before he reached his hide a fine boldly marked and brightly
coloured bird ran away from it. The black belly-patch, he noticed, was
strikingly larger and blacker than that of a rather dingy dotterel he had
just photographed. It was also a slightly larger and a noticeably plumper
bird. At 16.00 the mist came down and a steady drizzle wet his camera
lens. But Auger stayed in his hide until he was relieved.

'At 16.40 the sitting bird gave a short shrill call and sat very taut and
erect. Peering through the window of my hide I now saw a second
dotterel about 20 feet away. This bird ran towards the nest in short
bursts. When it was about 6 ft away the first bird called softly and the
sitting bird ran off and the new bird settled on the nest. It was still
sitting at 18.30 when I was relieved. Rain was falling steadily. I saw no
sign of the second dotterel from 16.40 onwards. The bird which had
relieved the bright bird on the nest was of identical dinginess with that
which I had previously photographed at the first nest.'

On 16th June, Auger sat in his hide from 15.30 to 17.30, mist again
making photography impossible. 'The brooding dotterel was definitely
the large brightly coloured bird and the dingy partner never turned up. I
do not know what stage incubation had reached, but none of the eggs
showed any sign of chipping.'

Sam van den Bos possibly saw a hen sitting in the east Cairngorms.
'The smartest dotterel I ever saw on a nest was joined on the feeding
ground and, once when calling after leaving a nest, by a much poorer
specimen.'

A few nineteenth century naturalists claim that they shot brooding
hens in Scandinavia. In 1884 Alfred Chapman shot a hen off a nest in
Sweden and in 1893 Pearson and Bidwell killed a sitting hen in Arctic
Norway. In *Beyond Petsora Eastwards*, Pearson also notes: 'Of seven
dotterels shot from the nest or standing beside broods, two were hens.'

On the other hand, Berg in Norway and Sweden and Barth in Norway, Franke in Austria, Sollie in Holland and Tewnion in Scotland found only cocks sitting.

Rittinghaus had a different experience. On 6th July, 1958, he ringed a sitting dotterel, but on 9th and 10th July found an unringed bird on the nest. At two other nests both partners sat. Here is one nest-diary.

At 16.10 on 2nd July, 1959, just before the hen laid her second egg, Rittinghaus trapped and colour-ringed both dotterels close to the nest.

At 14.55 on 5th July the cock was sitting on three eggs.

Between 6th and 8th July stormy weather prevented observations.

At 11.58 on 9th July, however, the hen was sitting and a second dotterel – probably her mate – was close to the nest. At 13.47 the hen was sitting, and the cock ran towards her. Both birds then flew away. Then, at 14.02, the cock came off the nest, but the hen was absent.

11th/12th July

22.23 hen brooding.

22.50 cock one and a half kilometres away, feeding with four other dotterels.

0.16 hen sitting.

02.37 hen sitting, cock with a party of eight other dotterels on the west flank of the hill.

03.46 hen sitting, cock present.

05.19 hen sitting.

06.20 cock arrived to relieve hen, which then flew away towards the feeding ground.

At 05.42 on 13th July, after the cock was flushed from the nest, he carried out distraction displays and at 09.20 he again did this.

The cock was brooding at 23.08 on the 17th and 01.15 and at 03.27 and 07.18 on the 18th. Stormy weather between 19th and 23rd July prevented further observations. But by 24th July a predator had apparently destroyed the eggs.

Rittinghaus and Pulliainen recorded that both dotterels had brood-patches but the only two breeding hens that I have examined lacked them. Reg Wagstaffe examined the series of skins in Liverpool Museum but found no brood-patches on either. Stage of incubation and individual variation possibly account for these differences. In Finland O. Hildén discovered that at least two hens brooded during the hatch. In 1965 both cock and hen sat on a nest with one chick and two chipped eggs. In 1966 the same colour-banded bird was incubating on 8th and 11th July but on the 11th an unbanded dotterel was injury-feigning near the nest. During the hatch, on 12th July, the unbanded bird brooded all the time.

Pulliainen (1971) recorded both adults incubating at three nests. In

1969 a cock and hen brooded alternately at the end of the incubation period. In 1970 another hen shared incubation during the first fifteen days, spending 10% of the period on the eggs. Towards the end of the incubation period only the cock was seen on the nest. The polyandrous hen on Värriötunturi Fell was as unusual as Blackie. At her first nest she brooded for short spells on two days while she was laying. During the following twenty-four days the cock alone incubated. At her second nest she brooded for 53% of the incubation period, the cock for only 47%. At five other nests the cock only was seen incubating.

These detailed nest-diaries complement observations in Scotland. What really is the hen's rôle in brooding? Early in incubation some hens clearly do stints of both day and night duty and at least two hens brooded during the hatch. All observers, nevertheless, agree that the cock's share in brooding greatly exceeds that of the hen. Polyandry and regular brooding by the hen are possibly incompatible, but Blackie actually laid a clutch for a second mate and then helped to brood it!

The pair at the nest

The two dotterels are seldom seen at nests containing eggs. Indeed, whenever Macomb and Edgar Chance saw a pair together, they assumed that they had no eggs. This is not, however, completely true. Apart from both sitting during egg-ceremonies or on incomplete clutches, both dotterels do sometimes brood eggs.

A hen also sometimes calls the cock off eggs and then escorts him to, and perhaps from, the feeding ground. In May 1929, Edgar Chance wrote: 'In appalling cold and sleet the hen rose and called off the cock, which got off the nest some 200 yards away.' In 1934, Blackie called her first mate off the nest and, in a snowstorm the same year, another hen answered the cock after I had flushed him from eggs. In June, 1940, I likewise watched a hen fly in to join a cock which had just left his eggs. Hens particularly approach nest and sitting mates early in the morning and in the evening.

In the central Cairngorms, on 3rd July, 1966, Tewnion saw a hen close to a nest. After the cock was flushed he ran over to the hen and then both birds ran about on the snow. The cock eventually returned to the eggs and 'kept *peeping* as if to reassure its unhatched chicks', then presumably near hatching. The hen now flew up and 'after a good half hour gradually drew nearer. However it did not come right up to the nest but halted about 100 feet away.' The chicks were still in the nest on 9th July.

In June, 1950, Tewnion 'noticed a dotterel by chance when looking for young ptarmigan. It was displaying, so I sat down and watched.

After about a quarter-of-an-hour it stopped running about and settled on its nest, on the bare moss only about 50 yards from where I had stopped. I walked very slowly towards it, but after a few steps I saw the hen about 10 yards beyond the nest. When I approached the nest the hen flew away and the cock also flew off while I was still about 10 yards from the nest.'

Van den Bos has had a rather different experience. 'When the sitting dotterel goes off to feed it is usually joined by its mate which occasionally chases it off the feeding ground to make it return to the nest. Once I saw the non-sitting bird accompany its mate right back to the nest.'

J. F. Peters watched a hen visit a bird which was brooding a hatching clutch. 'Both birds by the nest were bright and well-marked and the bird that was sitting called several times. When the bird was put off the nest, it ran towards the other and then flew at it; the other rose and flew right away until lost in the cloud. The other bird almost ran back to the nest.' It is just possible, however, that this second dotterel was a trespassing hen.

The two mates are thus seldom seen close to a nest containing a complete clutch of eggs; but this is not an inflexible rule.

Brooding rhythms

The brooding rhythms of dotterels greatly vary. In warm weather a cock sometimes feeds twice hourly, coming off the nest and then running over the hilltop or shoulders. Within ten minutes he then returns to his eggs. But this is not typical behaviour. In June, 1936, we watched a nest from 10.23 to 21.37. At 11.17 the cock took a quick meal lasting exactly seven minutes. He then sat from 11.24 until 13.45 when he suddenly flew off over the hill. Twenty-seven minutes later he flew back almost to the nest, and sat until 18.23 when he was away for twenty-nine minutes. Another dotterel now flew ahead of him, but did not land. The cock then continued to brood from 18.52 until 21.27.

Another cock watched from 08.11 until 16.37 on 26th June, 1942, came off at 11.37 for sixteen minutes and then sat until 15.31 when he was away for twenty-two minutes.

In severe weather, some dotterels undertake much longer brooding. On 21st June, 1934, snow fell in the night. At 20.30 on the 22nd I found that a cock dotterel was sitting on a nest completely surrounded by snow, on which there was no mark whatever. He had thus sat throughout the day and night without leaving the nest.

At a nest in the Cairngorms in which the eggs were hatching, Seton Gordon recorded that between 09.45 and 13.15 a cock twice left for meals which lasted five and thirty-two minutes. Two days previously he had left his eggs four times in two hours. These observations were recorded in hot weather.

Blackwood suggested that dotterels leave their eggs uncovered 'for hours at a time if it is very hot', but he gave no detailed observations. On 11th June, 1960, however, Auger watched a cock sitting for at least one and a half hours – 16.00–17.30 – and on 12th June for four and a quarter hours – 15.00–19.15. On both days the hen never came near the nest.

On 30th May, 1912, Peters watched a dotterel sitting on three fresh eggs. 'Only one bird there, nest uncovered twice, with two hours' interval, and flew to higher ground to feed. When returning lit some considerable distance from nest and ran directly to it.'

On 3rd July, 1969, Pulliainen intensively watched a dotterel's nest. The hen incubated from 12.08 to 15.24. The cock relieved her at 15.38 and continued to sit until 17.53 on 4th July when the hen took over at 17.57. The hen then brooded until 09.26 on 5th, and the cock replaced her at 09.27. The cock then alone incubated on 5th and 6th July. Cock and hen thus brooded alternately at the end of the incubation period, but the cock was on the eggs during the hatch.

Incubation period

Table 16 gives an approximate average brooding period of 26.1 days for seventeen clutches. The last egg appears to hatch in slightly less time than the first two eggs. The intermittent brooding of the first egg thus does not apparently advance incubation. 'The double-clutch of four eggs from the Cairngorms in 1968 evidently hatched after about 23 days.' (A. Tewnion.) In exceptionally warm summers a dotterel also sometimes hatches a replacement clutch in a shorter period than those shown in Table 11. The full brooding drive probably generates more rapidly in repeat clutches.

Behaviour at the nest

A brooding dotterel sometimes turns its egg with its beak. It rises on the nest and then pushes its head down into the cup. It also shuffles the eggs with its upper breast and belly, thus fitting them into position. Auger tells me that 'on two or three occasions the bird ruffled rather than turned the eggs and that each time that it settled down again it had possibly changed the position by ten to fifteen degrees. It eased itself over the eggs and did not stand up. This was possibly due to the strong wind, because the bird frequently flattened itself as if for shelter.'

Sitting dotterels also periodically toy or fidget with the nest lining or with pieces of moss or lichen on the rim.

Francis Nicholson noticed that a dotterel was brooding head to wind. 'It always sat on the egg with the small end towards the wind. The egg was not on the same position on the second occasion, and it struck me that it sat on the eggs so that it could have its head to the wind and move its head accordingly. The first time I found a nest the big end was pointing due north, from whence the wind was blowing, and the bird sat on the egg facing north. On a second visit, wind west, big end of egg and bird faced west.'

A nervous or frightened dotterel crouches over its eggs in a stiff posture, but when brooding is normal and relaxed, or particularly when the chicks are squeaking in the eggs, it sits with head and neck upright.

Sitting cocks sometimes allow the searcher to come within a few feet before they slip off and run away. Some sit so tightly that you almost tread upon them. Then, chittering plaintively, they carry out fantastic distraction displays. On 16th June, 1934, I suddenly saw a dotterel sitting less than a foot away, in a little hollow barely forty yards from a melting snowfield. As Carrie bent down the dotterel jumped off and flew up almost into her face. Then it dropped to ground, flapping its wings and chittering half-heartedly. The little fellow then limped towards us, presently stumbling and almost falling in the stones. Then he stopped abruptly and ran back to his eggs, snuggling down quite fearlessly. Kneeling beside the nest, Carrie first stroked and then gently lifted him and put him back on the nest again. The little cock shook himself for a few minutes and then squatted down on a rock about six yards away. For three minutes he was silent. Then he gave a shrill squeaky *wheep*, and *skirring* angrily he ran over the stones, picking up spiders and chasing craneflies. Ten minutes later he was brooding, but still treating us as if we were wandering red deer. When we walked away he left his eggs and ran after us with wings held above his back. We left him crouching on a rock and watching us.

In the early 1920s James Ramsay MacDonald and some of his family were caught in a thick mist and forced to spend a miserable night on this ridge, and so we called our little dotterel Ramsay.

A few other arctic waders use this behaviour to divert big harmless mammals from the nest. A cock surf bird did this at the first recorded nest in Alaska. When a mountain sheep was about to step on the nest 'the surf bird exploded right in the astonished animal's face'. (J. Dixon, 1929.)

On Jenny Lind Island hen white-rumped sandpipers regularly, and Baird's sandpipers occasionally, use these tactics. The white-rumped sandpiper, running back to the nest, stopped abruptly, 'then she leaped straight out at us and fluttered before our faces. Not infrequently the incubating bird flew directly from her eggs and fluttered before us as we approached.' (D. F. Parmelee.)

Few dotterels brood quite so closely or are so aggressive as little Ramsay. But he was not unique. We have fondled quite a few on their nests.

In their runs back to the nest, some birds occasionally employ brief distraction displays. Others make rather quick runs through their territories in stiff and upright posture with their tails closed, periodically stopping and starting. A few fly off even if they have continued to brood until you have almost trodden on them. Then off they go with downspread tails and rather slow, clumsy or hesitant wing beats. An odd bird flies away when you are up to two hundred yards or more away from its nest. No distraction tactics here, just a quick flight above ground level.

Derek Ratcliffe and John Mitchell watched how a dotterel behaved after it had lost its eggs. 'The dotterel continued to act oddly, running back gradually to a particular place near the fence and then flying away again from one to several hundred yards, only to repeat the whole process. After about an hour of this perplexing behaviour the dotterel had not settled on a nest, and finally it flew away and did not return. We walked over to the spot in which it had been so interested and found a spoiled nest with freshly smashed eggs in it. Presumably it had been raided by a fox or a shepherd's dog and perhaps only on that same day.'

Egg retrieving

I have watched a cock retrieve an egg which he had accidentally knocked out of the nest. He did this quite easily and quickly by standing up in the nest-cup and hooking back the egg with the underside of his bill. Rittinghaus also found that a dotterel rolled back eggs which he had placed on the edge of the nest.

I tested one cock dotterel which had two eggs in a nest-scoop on a steep hill shoulder by placing both eggs, one three inches and the other six inches, on the uphill side. On his return, calling almost continuously and hiccoughing nervously, the cock sat in the nest making bill movements as if arranging eggs. I allowed him five minutes, but he failed to cope. Then I replaced the eggs. At first the little cock nodded, 'hiccoughed' and called, and then he quickly wriggled the eggs into position, and was soon brooding them as if nothing had happened. Further experiments may show that individual dotterels vary greatly in ability to cope with egg retrieving problems. This has certainly been my experience with lapwings.

Hatching

I always love to sit beside a really tame dotterel whose eggs are hatching. I particularly remember that little gem of 1940 which Carrie had found sitting on a stony tableland. He allowed us gently to stroke him on the nest. On the 13th our little dotterel's eggs were starring. On the 16th, John Markham watched the first chick hatch out. Once the hen ran almost up to the nest but, 'skirring' angrily, the cock drove her away. All that day he sat helping his chicks to reach a world of warmth and softness.

On 28th June, 1939, Seton Gordon watched a nest in which one egg had started to star, but the chick came little nearer to birth. On the 29th all the eggs had chipped, but on the 30th they still had not hatched. On 1st July the chicks' beaks were showing through but the nest was wet with snow and rain. This time the hen was near. She chased the cock back to brood. On the afternoon of 2nd July two chicks were out, one dry, the other wet. In the night snow had fallen. It was a cold, wet and bitter day with the tops white with fresh snow. When Gordon reached the nest the cock was off and the chicks shivering. Then, as a flurry of hail and sleet flashed across the hill, the cock suddenly sprinted back. 'I had just time to replace the chicks. I was then two feet from the nest but he had no fear at all and looked at me as if to say "I got back just in time didn't I." Once the dotterel rose from his eggs and took a cranefly off my kilt as I sat beside him.'

Edvard Barth considers that weather and state of incubation influences the tameness of brooding dotterels. In cold weather the dotterel will sit so tightly that you can hardly move him away. On a cold day in 1953, only a few degrees above freezing, a cock covered his chicks, with the shells of all three eggs just two inches away. At first he injury feigned, then he recalled the chicks. In the end Mrs Barth lifted the whole pretty family a foot off the ground with the cock still sitting on them! Once, like our own Ramsay, he flew up and pecked her!

As soon as the chicks have started to squeak, the cock answers them, even if no chip or star yet shows on the eggs. He thus treats the hatching eggs as if they were already chicks and he does not turn them so frequently as before. The chipped or starred side of the egg is then usually turned upwards. Brooding cocks with hatching eggs, however, do not all behave in the same way. A few do not sit closely and take long to go back.

Chipping period

The chick is sometimes long in emerging from a 'pipped' egg. The relative strength of chick and eggshell probably partly accounts for great differences. In the Cairngorms and Grampians, eggs in eight different clutches took approximately 25, 46, 48, 60, 72 (2), 96, and 112 hours for the chicks to break free.

Hildén had similar experiences. 'The nests were not controlled daily but, judging from the following records, the chipping period is about four days.' On 8th July, 1966, two eggs had sprung. At 15.00 on the 12th none had hatched, but three hours later there were two chicks and one egg.

On 2nd July, 1966, two eggs in another nest had chipped; on the 5th the beaks of two chicks were protruding and the third egg was chipped. On the 6th two chicks had moved about fifty yards.

One egg in a Cairngorm clutch hatched in roughly twenty-five hours after it had first pipped, but this was not unique. At 18.30 on 17th June, 1943, George Yeates found a Drumochter nest where one egg was split in two, the second starred, and the third unmarked. At 20.00 on the 18th, all three had hatched. The cock was brooding them in the nest.

In 1949 Franke noticed that one egg in a clutch of three was 'holed'. By 07.00 the next morning one had hatched, the second chick emerging at 9 o'clock, the third at 10 o'clock; but we do not know whether the two later eggs had starred on the 24th. The chipping period was nevertheless brief.

I do not know whether chicks hatching after prolonged chipping periods survive less successfully. A golden plover chick, for example, which took over a hundred and twenty hours to leave the chipped eggshell, died within a few yards of the nest.

Hatching spread

All three chicks in a successful hatch usually hatch out in a spread of less than twelve hours; but in two clutches of three eggs there was a spread of over sixteen hours. The interval between the hatch of the second and third egg is usually longer than that between first and second. The third egg thus tends to have a slightly shorter brooding-period than that for the first two. This is probably caused by the cock not immediately generating the full drive and brooding temperature. Warmth from the bodies of the first two chicks also possibly assists the last egg to hatch rather more quickly.

The signals between parents and chicks during hatching probably have important survival value. By listening to the cries in the eggs, the brooding bird may be able to reduce spread. The chicks now also start to learn the parent's call-signals. The cries of older or stronger chicks also probably stimulate the hatching process of the younger or weaker, another adaptation to reduce spread and promote more synchronous hatching (cf. Bobwhite quail, M. A. Vince 1964; Stanley crane, L. H. Walkinshaw, 1963).

In 1949 Franke found two chicks and a chipped egg at 05.45. The third egg hatched at 11.30, at least six hours later than the second. In 1952 two chicks hatched within four hours, the third and last laid – a rough-shelled and sparsely-speckled egg – was infertile. The two chicks hatched in bitter weather during a 100 km. per hour gale and at an air temperature of 1° C.

At a nest in Finland there was an equally synchronous hatching. All three chicks hatching out on the same night (Hildén). On Värriötunturi, Pulliainen recorded a hatching-spread of 5 to 19.5 hours, with 3 to 6 hours between first and second chicks, and 2 to 13.5 hours between the second and third.

Eggshell disposal

After the chicks have hatched, the cock dotterel usually removes the empty eggshells. After pushing or lifting large pieces on to the rim of the nest-cup, he usually flies away with them, dropping them twenty to twenty-five yards from the nest. One cock, however, flew low and fast for about two hundred yards before releasing a shell. We watched another cock fly off with a large chunk of the third egg to a small tarn about fifty yards away, where he dropped it and then rinsed his bill. Sometimes, instead of dropping the eggshell, the cock lands, breaks it up, and then eats small bits. I once saw a cock peck a piece of eggshell into small fragments and then poke them in a patch of *Rhacomitrium*.

John Markham photographed an exceptionally fearless cock dealing with a large fragment of the first egg. By body movements he shuffled the eggshell three inches from the nest. Then, for over an hour, he continued to brood, apparently ignoring it. At last he leaned over and started to peck it. Exactly 100 minutes after the chicks had hatched, the cock lifted the shell and flew away. He was off the nest for twenty seconds. On return, he immediately picked up another shell, ran about twenty yards, and then pecked and nibbled it. This time he was away for two minutes. In 1945, on the other hand, a brooding cock acted more promptly, flying off with the shell immediately the chick had left it.

Seton Gordon had a different experience. The father dotterel waited until all three chicks had hatched before dropping the shells, one after the other, about twenty-five to fifty yards from the nest.

Franke filmed a dotterel pecking at the shell while the chick was coming out. Later, he carried the shells about twenty yards and again pecked them and possibly ate small pieces. A dotterel is seldom lax about removing its eggshells. In 1968, however, Adam Watson examined the nest in which the two hens had laid four eggs. The chicks had left but all four shells were still in the scoop, and one was tucked hard inside the broad end of a second, and the broad end of a third inside the sharp end of another. The chicks had neatly cut circles round the eggs nearer to the broad end than to the sharp end. The entire remains of all *four* eggs were still in the nest-scoop.

Dotterels eat, push under the nest-lining, or completely ignore small fragments.

We placed large and small bits of a domestic fowl's eggs in a dotterel's nest with three highly incubated eggs. At first the cock ignored these foreign bodies. Then, one by one, he flew away with the two large pieces, dropping them about twenty yards away. Each time he flew into the wind and fluttered just before dropping the shell. A third fragment he carried uphill and partly ate. He also ate most of the small pieces.

What is the point of this behaviour? The removal of eggshells is clearly pro-cryptic. A hatched eggshell, with the membrane lying upwards, close to the nest, lessens the cryptic value of the colour-patterns of the eggs, chicks and parent. An eggshell lying in the open is thus an easy mark for a crow, gull or skua patrolling overhead.

Natural selection has evolved eggshell disposal as a means of passive defence, but many complex factors are involved. A dotterel deals differently with large and small shell fragments; probably due to balance between conflicting risks. A large chunk of shell with a larger exposure of white membrane is more likely to catch a predator's eye. Large shell-splinters also possibly irritate the parent's brood-patches or cut tender-skinned chicks. The brooding bird thus risks flying from the nest. Yet, unlike many other waders, a dotterel seldom flies far before dropping its eggshells.

Is this behaviour a result of natural selection? Here are some relevant facts.

(a) Short flights lessen the chances of detection by predators.

(b) Brief absences from the nest ensure that the chick suffers minimum exposure. In severe weather dotterels sometimes allow egg-shells to pile up in and around the nest.

(c) Lacking the co-operation of the hen, the cock must prevent older and stronger chicks prematurely straying. These different hazards probably explain why a dotterel so seldom flies away with small pieces. The advantages of removal are unlikely to compensate for other risks.

Why, then, do not all cock dotterels quickly or immediately remove the hatched shells? Possibly because brooding and chick-tending – two different drives – are in conflict, thus causing an emotional crisis. Some dotterels adapt more quickly to the change. I do not know, however, whether those which react slowly survive less successfully.

Natural selection has achieved a bewildering variety of eggshell disposal patterns, each presumably geared to particular needs. Let me examine a few examples. Although dotterel and ptarmigan often share breeding-habitats, they have different methods of coping with discarded shells. Dotterels usually fly away with large fragments whereas the hen ptarmigan probably eats a few bits, but leaves most of the eggshells in a heap in the cup or on the rim. Often, however, she pokes some caps into the main portions. I cannot suggest a satisfactory explanation of this difference, but a hen ptarmigan is better able to defend her chicks than a dotterel, and less frequently permits hatching eggs and small chicks to remain unbrooded. Ptarmigan chicks also usually stay a shorter period in the nest than do dotterel chicks. Hen ptarmigan and red grouse, however, sometimes rake in and then continue to brood large shell fragments which have accidentally rolled out of the nest during the hatch. This is presumably pro-cryptic behaviour.

Gamebirds (*Galli*) all lay large clutches. During incubation the close-sitting hen ptarmigan's cryptic mantle helps to conceal her and the eggs from predators. Removal of the many eggshells would involve the hen in numerous journeys and subject small chicks to exposure. Movements to carry away eggshells would also probably betray the brood to predators. For gamebirds, therefore, greater risks offset the gain of eggshell removal. Natural selection has thus acted decisively against active disposal. Almost infinite natural experiments gradually eliminate more dangerous expedients.

Many facets of the shell disposal techniques of waders need investigation. Why, for instance, do dotterels and others sometimes drop shells in water and then rinse their bills? Is shell-floating a throw-back to periods when waders possibly nested in wetter biotopes? Or do predators tend to ignore shells floating on pools or water-holes? What is the survival value of beak-rinsing? Is this behaviour comparable to that of the dipper which drops its faeces in a stream and then washes its beak before collecting more food for its chicks? Why do waders deal with small shell fragments by so many different means? Individuals as well as species vary so greatly. Yet lapwings, for example, regularly hide little bits of hatched shell under the lining. This behaviour (which some dotterels share) must have survival value.

Is pecking and breaking up a hatched eggshell really a re-directed attack? A cock stone curlew wrenched the tops off both eggs to help the chicks to free themselves and then continued to brood for about forty-

five minutes, with a chunk of shell held in his bill. Later, he deliberately
stamped upon, broke up and ate this bit of eggshell. At another nest,
the cock carried away one large portion of shell and then pecked it to
small pieces. After this he ran back to the nest, but later returned and
ate all the fragments which he had no difficulty in re-finding. (E.
Hosking, *pers. comm.*)

Tinbergen (1962) learnt that robber gulls preferred to eat wet,
rather than dried-off, nestlings. Has this peculiar eggshell behaviour
evolved to inhibit parents attacking their own chicks?

The complex of hatching certainly poses endless problems.

CHAPTER 11

Young Dotterels

Dotterel chicks and behaviour in nest

DOTTEREL chicks are small and lovely balls of tortoiseshell fluff, with down patterns containing subtly different shades of brown, ochre-brown, black and buff, mixed with various tints of black and white. The broad white superciliary stripe and the white ruff-like crescent extending from chin to chin are diagnostic. When it first leaves the egg the chick's bill is lead-grey, but this colour soon darkens into black. At first its legs are lead-grey, but there are yellow patches on the joints and on the soles of its feet. Its claws are brown with grey tips. The irides are dark liquid brown.

How futile to try to describe the wonderful grace and beauty of any wader chick!

If you disturb a cock dotterel whose chicks are still in the nest, he runs away looking almost like a small mammal; then perhaps he stands still, making nodding or hiccoughing flight-intention movements. After he has gone back to the nest, he again sits rather high and continues to jerk or nod his head. The chicks below him give subdued *kees* which he answers or imitates with almost equally soft *kwee kwees*. Time after time he squats and shuffles over the little bundles of fluff and bends forward to arrange or re-group them.

In less than six hours after hatching many chicks are strong and active enough to leave the nest and totter away from their father's warmth and attention, but he usually recalls them, seldom allowing them to stray far, as in the nest they presumably grow stronger, as do young poultry, grouse or pheasants. In cold and stormy weather the father broods his chicks in the nest for up to twenty-four hours. The rich source of food in the yolk-sac inside them now enables them to survive. I have already described why confinement to the nest is also likely to have other survival values.

Here are a couple of records of dotterels with chicks. At 3.15 on 8th June, 1940, we found a cock brooding two chicks in the nest. One was wet and still weak, but the second tottered a few feet and then crouched down beside a stone. At 21.39, however, the cock was again brooding both chicks in the scoop. Presumably he continued to do this throughout that night. The next day, at 19.15, another cock was brooding a nest with a trio of chicks, the youngest just hatched. We watched the father dotterel fly away with a chunk of eggshell. At 08.14 the next morning all three chicks were still in the cup. Yet these were warm summer days.

In 1949 Franke found that two chicks dried off in two hours and six hours later were trying to wander away. The next day, however, the cock was still brooding the trio in the nest-scoop. At another, Franke found a chick about a yard from its nest. It then ran back to the scrape where the cock was brooding its two siblings six to eight hours old.

At three nests Pulliainen records that the youngest chicks were 10.7 to 14.2 and the oldest 19.5 to 30 hours old when they quit their nests. The youngest chicks were then rather weak and tottery. The broods left during the morning and afternoon.

The nest-scoop protects small chicks from the wind. Many chicks nevertheless die in early days. Weaklings probably fail to keep up with their brothers and sisters and thus lose their father's protection. In severe storms or in heavy rain, or in showers of hail, the cock periodically broods his family. The survival advantages already described partly compensate for the lack of a second parent.

Share of sexes in parental care

The cock usually alone tends the young broods, but four times in the late afternoon and evening I have watched a hen – presumably the mother – approach cock and chicks. Each time the cock drove away the hen. Nevertheless, some cocks do accept hens and allow them to run with the broods. George Yeates watched a pair escorting three chicks. 'The hen took no part in tending the group, but seemed more like a foreman or supervisor. She made short flights just ahead of cock and chicks.' The chicks were then one week old. This observation is particularly interesting as its locus was a hill where several pairs nest annually. Pulliainen made several new discoveries. Some cocks accepted or adopted chicks from other broods but these waifs sometimes returned to their own fathers. One cock attacked a small chick from a strange brood. Pulliainen also watched fights between cocks with broods and strange dotterels, probably their own hens.

Isolated pairs more frequently accompany broods, possibly because the absence of other hens prevents grouping. On a large Atlantic island off Norway, where dotterels are few, K. Avset located two broods; with one of which both parents were running.

In Lakeland both cocks and hens sometimes associate with the broods. In 1917, for example, George Bolam watched a pair with three one-day-old chicks and in 1921 Bolam and G. W. Temperley also found a pair escorting small chicks. On another fell J. F. Peters saw two old dotterels with a late brood of three to four day-old-chicks.

In the west Grampians, in 1962, A. Watson, Sen., also watched a pair with one well-fledged youngster, and on 18th July, 1971, P. Stirling-Aird had the same experience in west Perthshire.

In 1966 Hildén watched a pair escorting a brood two hundred yards from their birth-scoop and in 1952 Franke found that the polyandrous hen left her second mate sitting and then returned to her first mate and brood. For several days the cock fiercely drove her away, but finally accepted her when the young were almost fully fledged. The social or flocking impulse was then presumably dominant.

Rittinghaus watched strange dotterels approach cocks and broods. Two larger adults – probably hens – joined cock RY – one actually attacked him. But he associated with the second – possibly the mother of his brood – allowing her to seek food close to the foraging young. While tending a two-week-old brood, cock YB also moved about with two adults, one certainly a hen. Cock WW drove off most adults, only allowing one particular hen – possibly his mate – to approach his youngsters. Rittinghaus also watched two adults accompanying flying young. At first the adults flew off, then the cock came back and the hen soon started calling and then slowly worked uphill with a third young dotterel. All five birds then joined up in a loose group. In August, 1958, Rittinghaus also watched two adults running beside a juvenile. All three birds belonged to a trip of dotterels, consisting of six young and three old birds.

In the Grampians, Ratcliffe likewise met with a small family group, consisting of cock and hen and three fledged young, and on 31st July, 1970, four of us watched a pair with one flying youngster in the east Cairngorms.

Family parties are thus sometimes integral parts of pre-migratory groups in summer.

Chick mortality

Table 17 gives some data on chick mortality, but we do not yet know what proportion die because of inherent weaknesses already present in the egg opposed to severe weather or to shortage of food.

Recent research suggests that the quality of the egg is certainly an important factor in determining breeding-success in red grouse and possibly ptarmigan. 'One possibility is that the production of young (in Scotland) or their survival rate after summer (Iceland) is affected by factors after the eggs hatch, for instance by bad weather, lack of food, or food of poor nutritive value, which might all weaken or kill the chicks.

Alternatively, the survival of the chicks might be pre-determined before they hatch by the quality of the eggs and in turn by the nutrition condition and possibly "quality" of the parents.' (Watson and others, 1969.)

In the dotterel, however, only the first hypothesis is likely to apply to conditions in Scotland. The second presupposes some failure of food on migration or on the winter grounds. There is, nevertheless, already some evidence that young dotterels usually thrive better on the more richly vegetated and presumably insect-richer hills in the east Grampians than on the desolate granite barrens of the central Cairngorms.

Once the chicks have used up the food reserve in their yolk-sacs their survival largely depends on the cock's ability to herd them and lead them to food. In the Cairngorms early broods sometimes unexpectedly survive heavy summer snowstorms (cf. p. 60) but sleet, hail and long spells of high east wind are probably greater hazards. I have no direct proof, but I twice noticed that a chick was missing after a sleet storm. On 2nd July, 1947, moreover, Watson found a dead dotterel chick at 2,250 feet near the path south of the Wells of Dee. A high wind had probably blown it off a slope of Ben MacDhui.

Care of chicks

In their first few days, the young dotterels seldom move far from the nesting territory. Thenceforward, they move varying distances, but broods usually remain apart and separated because the cocks fight and challenge one another. Rittinghaus suggests, however, that the broods sometimes travel many kilometres with their parents and that they pass over different levels on the hill. In the Cairngorms broods appear to make shorter treks.

At first the chicks usually crouch when the cock gives warning calls. Sometimes, however, quite small chicks tire of lying still and stand up and run away.

Seton Gordon tells me that he has seen chicks when only one day old 'bow' in true dotterel fashion.

The cock periodically broods his chicks during the day as well as at night; but small chicks can survive for several hours without brooding. The cock, however, intermittently broods his chicks well after they are six days old. In 1945, for example, we watched two broods aged about nine and seventeen days. The cock frequently brooded the two nine-day-old chicks and only commenced distraction displays after he had returned to brood them. For the previous half-hour he made short flights up and down the hill, much as a golden plover behaves when its unfledged young are in the moss. The chicks, meanwhile, had probably become cold and chilled. The second cock included a display of false scrape-making as one of its distraction activities.

We carefully examined the young dotterels. The legs of the nine-day-old chicks now had yellow calves, but their shins were still dark leaden-grey. The legs of the chicks of the older brood, however, were now almost completely yellow – only faint grey streaks showing on its shins. The younger brood did not crouch when the cock called, but ran away quickly. On the other hand, the older chick skilfully hid itself in tussocks of grass. Once concealed, it was difficult to find, so well did its down-pattern harmonise with the herbage.

When the chicks are older, and able to hide themselves, they often run and then crouch in a tussock. The cock now flies off long before you reach him. Previously he had tried to distract you, running over the stony or mossy carpet, calling quietly or feigning injury. Now he makes short circular flights, dropping down, running a few yards, then flying backwards and forwards and often calling excitedly.

The wings of the young dotterel slowly grow and strengthen. One day it runs fast in front of you, half-flapping its wings to give it speed. On the next, perhaps, it flutters a few yards before flopping down, then it runs on or perhaps tries to fly again. Or, quite unexpectedly, it raises and spreads its wings and flies heavily away. A few days after its first faltering flight the young dotterel flies almost as strongly as its parents.

The period from chick to first-flight is always difficult to measure as it is almost impossible to determine when the youngster is really capable of flying. Three young dotterels from different broods flew strongly at some time between their twenty-sixth and thirtieth days, when they were already fully grown and, in the field, indistinguishable from adults in body-size. Young dotterels, however, are able to flutter and take short flights fully a week before they are capable of continuous flight.

Departure of young and old dotterels

From about the second week of August onwards, dotterel trips or family groups begin to leave the high tops. In late summers like those of 1951 and 1955, I watched mixed parties of old and young in the high Cairngorms in the last week of August. On 29th August, 1951, Adam Watson watched a solitary adult on the Yellow Moss of Mar in Aberdeenshire and on 7th September, 1947, he and V. C. Wynne-Edwards saw a trip of six at the Wells of Dee on Braeriach, where Wynne-Edwards had already seen a party of seven on the plateau (4,149 feet) in the last week of August.

In the earliest summers the entire dotterel populations has left by the end of the third week in August. On 18th August, 1967, in the central Cairngorms, for example, Watson recorded a flock of ten, a single adult, and a pair with one fully grown juvenile with slight touches of down on the sides of its head. On 20th August Bob Moss worked this ground but met with no dotterels. In the east Cairngorms in 1967 one solitary adult was the only dotterel remaining on one particular hill on 10th August, one juvenile on another hill on 13th August, and none on either hill on 23rd August.

On 10th and 11th August, 1967, only one solitary juvenile remained on an east Grampian top where trips had been seen earlier. By 20th August there were no dotterels on the hill. (A. Watson *in litt.*).

In the east Grampians dotterels probably nest a little earlier and their broods grow and develop rather faster than in the high Cairngorms, possibly accounting for slightly earlier movement of groups and trips. On 8th August, 1965, for example, there were seven old and nine young on a good dotterel hill in the east Grampians; by 21st August all had gone. On 14th August, 1966, the Watsons and H. G. Lumsden searched the summit plateau and all ridges above the 3,000 foot contour, where they located three groups of dotterels. The first consisted of six young, two old and two golden plovers. In the second party were two young, seven old, and on another ridge was a single juvenile on its own. On 16th and 17th August A. Watson, Sen., and H. G. Lumsden again hunted this hill but found that all the dotterels had gone.

I have insufficient data to compare flock and departure patterns in the central Grampians. On 21st August, 1953, we could locate no dotterels on a range of hills where at least eleven pairs had nested during the summer. But in 1936 on another hill group there, the dotterels stayed long enough to allow a party of sportsmen to shoot some while out after golden plovers in the third week of August.

Young dotterels sometimes stay behind after the trips have flown away. On 23rd August, 1951, for example, I watched a fully fledged youngster on a central Cairngorm hill from which all others appeared to have gone. It uttered the usual adult *whirring* anger calls. On 14th August, 1966, and on 11th and 12th August, 1967, Watson also recorded solitary juveniles as the last dotterels seen on the east Grampians. Young and old of the earlier broods usually join with others to form mixed trips, but the two parents sometimes accompany juveniles of late broods.

There are a few observations from abroad. On 31st August, 1952, Franke watched a trip of eleven dotterels flying south over nesting grounds in the Austrian Alps, but he naturally did not know whether these were migrants from Lapland or breeding birds on the move.

In Sweden Rittinghaus mentions that the earliest young on the wing were seen on 3rd August, 1958, 5th August, 1959, 24th July, 1960, and 1st August, 1961. Towards the end of the nesting season the dotterels in Abisko formed flocks or parties of up to thirty or more. These groups took wing at increasing distances from man. On 1st August, 1959, he located one of his colour-ringed pairs in a group of twenty-five to twenty-seven dotterels. About 20th August, 1967, Adam Watson also saw several dotterels on their breeding grounds in Finnish Lapland, but in 1969 on Värriötunturi Fell flocks did not assemble until mid-August.

In the U.S.S.R. the dispersal pattern possibly differs. In the last third of July a few hens appear on tundras in the Yenesei Gulf where they have not nested. About ten days later trips of cocks arrive and juveniles a little later (Tugarinov and Buturlin, 1911).

Some other wader chicks in the nest

Species like the dotterel, which have evolved single-parent fledgling patterns, are at an advantage if the older young stay in the nest until the youngest or weakest is strong enough to leave. The yolk-sac inside the chick has to be large enough to maintain them until they are able to respond to parental signals and to seek their own food. This pattern is less essential when both parents are participants. One parent can then lead away the older chicks while its partner stays behind to brood the youngest or weakest in the nest. The American and European golden plovers, *P. dominica* and *P. apricaria*, and the grey plover, *S. squatarola*, generally produce chicks which remain longer in the nest than those of almost any other waders. These chicks are hatched from eggs which are proportionately larger than those of all other Charadriform birds (Lack, 1968) and are usually hatched after exceptionally protracted 'chipping periods'. Both parents, however, tend the young, each looking after a part of the family. Dotterels and golden plovers, however, rely more on their cryptic colour-patterns and distraction displays than in attacking and driving off predators. Presumably, therefore, the stronger the chick is when it leaves the nest the more probable is its survival.

Wader species which, alone or in groups, defend their chicks by attacking predators, those lacking cryptic plumage, or both, often have extremely active chicks which tend to leave their nests within a few hours of hatching. The chicks of avocets, lapwings and oystercatchers are all precocious, although if the last egg hatches in the evening, the hen usually restricts the whole brood to the nest before leading them away on the next day. Some obviously closely related birds, however, have evolved strikingly different nestling adaptations which differing ecology and behaviour partly explain. Redshank chicks, for example, usually stay considerably longer in the nest than greenshank chicks. Why the difference? Greenshanks nest on well-dispersed territories, relying on close brooding and their cryptic plumage for concealing their eggs. Aggressive and effective defenders, they often immediately lead their chicks from the nest to squashy places or to the shores of distant lochs. During these long journeys the small chicks sometimes have to swim rivers.

Redshanks, on the contrary, often breed in loose groups, are usually light sitters, and nest in higher herbage. Relying on noisy collective mobbing as a deterrent to predators, they seldom move their young broods so far and so soon as do greenshanks.

A greenshank lays a proportionately larger egg than a redshank, possibly an adaptation to provide the chick with a greater food reserve to help it make these obviously exacting journeys from nest to feeding ground.

The dispersion and nestling patterns of American golden plover and ruddy turnstone in the Canadian Arctic also makes an interesting comparison. The golden plover, like dotterel, breed in scattered pairs, roughly one pair to the square mile, whereas the turnstones, nesting on roughly similar terrain, form small nesting groups, each consisting of three to four pairs. Golden plover chicks stayed in the nest for twenty-four hours or more. Turnstone chicks, on the contrary, sometimes remained in theirs less than twelve hours. Both golden plovers and turnstones led away their broods from the nesting ground, but each pair of golden plovers tended to rear its young, separated from its fellows, in hummocky wet tundra ground, whereas the turnstones often brought up their broods in noisy, collective and aggressive groups. The turnstones were quicker to attack predators and better able to guard their chicks, whereas the golden plovers greatly depended on their chicks' cryptic down-patterns. The long nestling period of golden plover chicks is thus possibly an important adaptation to enable the chicks to grow stronger and better able to disperse while moving from nesting to fledgling biotopes.

Table 18 compares the dotterel's nestling patterns with those of some other waders, particularly those studied by D. F. Parmelee in the Canadian Arctic.

I can think of few more exciting and challenging research projects than to investigate and compare the patterns and assess the survival values of the many contrasting nestling and chick adaptations of waders. But an understanding of the behaviour and ecology of the parents will help us to solve many of the problems.

CHAPTER 12

Voice

THE music of waders always stirs and excites me. I have only to close my eyes and I am at home again in the grey wilderness of the north. Above me, now, a cock greenshank dances like a gnat high above the tarns and gneiss of west Sutherland. Or, rasping and growling, he lifts and flaps his wings as he slowly goosesteps up to his mate. I hear golden plovers wailing above a snow-packed moor or crying together in mournful evening chorus. I listen to the hammering tinsmith scolding of wood sandpipers with a brood or one or both yodelling richly above the nest on the flow. Or, I hear the wing-music of lapwings in the gloaming. All these sounds are thrilling, evocative and acutely nostalgic.

The voice of the dotterel equally belongs. No one who has lain in a tent at night in the misty hills ever forgets the little drips of tinkling sound, always beckoning and often tantalising.

Listening to trips of dotterels on the tundras of Yenesei, Maud Haviland described the beauty of their calls. 'Here, walking over the rough hummocky ground, we would suddenly hear a tinkle of notes, very soft and liquid, like the drip, not of water, but of something slower and richer – nectar perhaps – which, as it was the drink of the World in the Celestial Childhood of Things, surely must have been golden and sticky.'

Equally lovely was her rendering of the dotterel's cries at the nest. 'A little sound as inarticulate as a sob or a sigh but seemed to be wrung from the bird by the strength of her distress.'

I have never heard the voice of the dotterel quite so romantically, but all its cries have meaning to me. But the joy of listening is no longer enough. We have to analyse and interpret. In time midget tape-recorders will help us to give more objective descriptions. But, meanwhile, the taping of bird sounds is still an art in its infancy. On most days wind and rain still prevent us recording the hill birds. The mountain gods seldom smile upon the bird recordist; and when we play them back the revolving tapes produce a weird cacophony of hiss and wind-bump.

Tape recording of the complete vocabulary or language of the dotterel is thus likely to remain a counsel of perfection. For long we shall still rely on verbal renderings. Let me then try to describe the principal calls and cries.

Social contact calls

While feeding, a group of dotterels keep in touch with one another by soft *kwip kwip* cries. In Sweden Rittinghaus records group contact calls 'very soft and drawn-out *pue pue* or *pur pur*'. In the Cairngorms, in late summer, I have recorded other flocking notes. A subdued *wā-wā-* and in flight *ter-tee* and *ter-too*. I have heard these calls shortly before the groups flew south.

Flight and movement calls

In flight, cocks and hens often continuously call soft but far-carrying *peep-peep-peep*. The dotterels often give a brief burst of these when rising and flying away. Rittinghaus emphasises that the birds call in quicker tempo when they first rise, or when turning, than while flying straight ahead. He renders the take-off calls as a soft *pioor pioor* and suggests that the touch down calls are almost identical with ō sounds accentuated.

Alarm notes

Mildly disturbed, one or a pair often fly off giving soft tinkling or 'dripping' cries in rising pitch *wit-wit-wit wita-wita-wita-wee*. This is their usual reaction to remote danger, as when a man appears on the skyline of a distant ridge.

This call is similar to the contact note, but birds accentuate the second syllable. This call is not chattered or long drawn out, or particularly penetrating (Rittinghaus).

More urgent danger makes a dotterel rise and fly and he then uses a softer and less angry *trill* than that of a dunlin.

Alarm calls

While standing beside a stone or peering over the skyline a dotterel often utters a creaking call expressing mild alarm or anxiety. Blackwood aptly described this as 'resembling the regular creak of an unoiled wheelbarrow'. This is a most evocative cry.

A single, sharp, explosive, but rather thin *ting* is another contact call used in anxiety. A cock sometimes *tings* between or during nest-dances. Dotterels sometimes *ting* explosively when on territory or perhaps before steady brooding has started. A cock flushed from eggs also occasionally *tings* when standing close to the nest. In other anxiety situations members of groups likewise sometimes *ting* quite excitedly

The *ting* call is usually employed by a bird on the ground, but P. A. Rayfield heard a trip *tinging* as they flew overhead at night. I have also heard an excited dotterel *tinging* while flying low over the nesting territory.

All-clear calls

Rittinghaus describes an 'all clear call' *puerr*, which it often gives while ruffling its feathers.

Anger calls

When you raise a dotterel, particularly at its nest, it sometimes flies away uttering a sharp *skirr*. If it sees a possible predator close to nest and sitting mate, a flying dotterel also *skirrs* angrily as it passes over.

In sexual and territorial fighting, this harsh call becomes a louder, angrier, and more urgent *trill*. Perry described the reeling of a party fighting as 'similar to that of turnstones fighting on an English beach'. Giving this call as *dru* Rittinghaus describes a dotterel holding its wings up and excitedly calling a much louder *druuu*.

Sex calls

During nest dances dotterels use special calls. Soft, rippling or tittering *tee-hee-hee* wire-twanging sounds. Our birds also now use sharp *ting* calls.

In ground displays either cock or hen sometimes sings a sweet jumble of almost linnet-like notes. In 1834 Thomas Heysham wrote: 'A few would occasionally toy with one another, at the same time uttering a few low notes which had some resemblance of those of the common linnet.' Perry also recorded four dotterels persistently giving 'a thin linnet-like note, pitched so high as to be almost inaudible'. I particularly remember how Meeson sang this song after he had stepped out of the scoop in Sheila's favour. Dotterels apparently lack formal copulation cries, but they often mate during nest-dances in which those distinctive rippling calls play a part. They thus differ from greenshanks, lapwings and oystercatchers, which have distinctive mating or copulation calls.

In their 'winnow-glide' display-flights, hen dotterels use quick rhythmic sequences of the basic *peep* notes as a rudimentary song. This song carries a great distance but, to the human ear, is often difficult to locate. It reminds me of cock crossbills which employ rapid sequences of *chip* calls as one of several songs.

Calls at the nest

Dotterels use different calls at the nest. The cock, Ramsay, hissed when he 'exploded' off his eggs. We have also recorded many other calls of anger and excitement, including a soft whistling *wheep*, a quiet but explosive *toop toop*, a strident whistle *kwee*, an angry rattling *tirra*, and a very excitable combination call – *kwee-kwee-skirr*, as well as the harsh *skirr* just described.

The cock's main distraction call when fluttering with beating wings and drooped tail is a continuous chittering *kweer-kweerik-kweerik*.

Dotterels also chitter during distraction displays while leading small chicks. Rittinghaus heard a cock warn his brood with these *fuer fuer* cries when a man suddenly appeared out of a dense mist.

A dotterel does not continuously give distraction calls and displays, but periodically runs back or stands hiccoughing and calling softly *kee* or *kee-kee-ee*.

Possible change-over calls

At the exchange which Harold Auger watched, 'the sitting bird gave a short shrill call and sat very taut and erect'. Meanwhile, calling softly, the relieving bird ran up to the nest in short bursts of running. A pair which I watched frequently interchanging at the nest in July, 1953, continuously repeated soft twittering twanging cries, sounding like loose fence-wires in the wind.

Calls with young

A brooding dotterel continuously gives soft cries to the chicks squeaking in the eggs. These calls possibly imitate those of the chicks themselves. After hatching the chick's cry strengthens and the cock now warns it with a soft *whee whee*.

To a bird with nidifugous chicks and a single-parent pattern, cries in the eggs probably have important survival value.

(a) The cries of older or stronger chicks probably stimulate those younger and weaker (cf. p. 101).

(b) These calls are a bond between parent and chicks, and help transition from brooding to chick-tending.

A little later, when the cock is leading, tending and brooding small chicks, he continuously calls *kee-kee* to which the chicks answer with shrill *cree-crees*.

At first chicks often ignore the cock's warning cries, but they learn to distinguish between more and less urgent cries. When I lifted up a chick about a fortnight old, it gave strident cries, *kweea kweea*, not unlike the parent's distraction calls. Meanwhile, the cock chittered a few yards away, ending a sequence of distraction calls with soft *kip-kips*. A chick which Rittinghaus picked up also gave a soft *yr yr*, presumably the same as I heard.

A young dotterel uses an adult contact call when it starts to fly. Flying young also use *whirring* anger calls alone, separated from the brood, or with the flock. Perry describes a juvenile making 'short flights from one spot to another on the gravel with a repetitive musical *tru-u-u*'.

Calls in winter and on migration

On migration and in their winter quarters, dotterels seem to use the same social-contact and flight calls as in summer.

In this summary I have described all the calls that I have recorded. To the human ear the dotterel's vocabulary is less complex than that of many waders, but it admirably serves the needs of an animal living in bare windswept country. I look forward to good tape recordings of all basic calls. Spectographs will then probably teach us something of the dotterel's origins. For example, do the dotterel's calls have affinities with those of the caspian plover on the remote steppes and saltmarshes? Do the waders of the great barrens evolve sounds which carry great distances? The language of mountain and tundra birds will provide an exciting research project.

Every bird probably possesses subtly distinctive voice characters. By this means we may soon expect to identify individual birds with certainty. Such a technique will be a wonderful research tool. Signals between parents and chicks are another study in their own right. Young dotterel hunters in coming generations will thus make discoveries which will far outshine those of older naturalists.

CHAPTER 13

Dotterel Populations

DOTTERELS seem to have a weak *ortstreue* and are sometimes less firmly rooted to particular territories or hill groups than, for example, are greenshanks and lapwings. We followed up five hens, however, by means of their peculiarly coloured and distinctly shaped eggs. We traced one hen – the famous Blackie – for five years, one for four years, two for at least three years, and one for two years.

Blackwood likewise mentions finding nests in different years 'within a very small radius of a particular spot'. In 1911, for example, he located a nest within fifty yards of one that he had previously found in 1908. 'It contained 3 eggs of the same shape and markings as the 1908 eggs.' The markings on the eggs that he found in consecutive years were 'in every year of a similar type. I am quite certain that it was the same bird which nested there each year.' On the other hand, the two nests which were found on English fells in 1959 and 1960 and in 1968 and 1969 contained clutches by four different hens.

The evidence of ringing and colour-banding likewise points to a weak *ortstreue*. At Abisko between 1958 and 1961 Rittinghaus banded twelve adults and nineteen young and in Finnish Lapland in 1966–7 Hildén trapped and ringed seven adults at their nests. In subsequent years, however, not a single one of these banded birds appears to have returned to its nesting fell.

Dotterels have possibly failed to evolve a stronger homing instinct because the value of particular biotopes in the Arctic varies from year to year. The amount of snow or the lateness or absence of particular insects might easily reduce the value or attraction of previously favourable habitats. Flexible populations are thus possibly best adapted to exploit impermanency. We may also later discover that a comparatively large mobile floating population – possibly containing many yearlings – periodically fills gaps or restores waning populations. In the Cairngorms, however, there are possibly some fairly distinct and separate breeding populations. All five hens traced by means of their eggs always nested on the same hill groups.

Numbers

How many pairs now nest in Britain in most summers? Have numbers greatly changed in the last fifty or hundred years? Do particular summers produce especially high or low populations of nesting dotterels? What do we know about breeding density?

In late springs, with exceptionally deep and heavy snowcover, do dotterels delay nesting or do they shift to lower and less snowy tops? Do their nesting groups maintain a non-breeding surplus?

North of Great Glen

Acceptable nesting records are few. Between 1945 and 1971 I have summer records of dotterels on nineteen suitable hills, three in Sutherland, ten in Ross and six in Inverness. I can only guess that about a dozen or fewer pairs nest on these hills in most summers. I do believe, however, that during this cooler phase of climate, more pairs are likely to breed.

Monadhliath

In most years the numbers of pairs breeding in the Monadhliaths is probably less – say – half that for the wild country north of Great Glen. I thus mark the Monadhliaths as half-a-dozen pairs.

Cairngorms

In 1885 'a capable observer' saw at least five pairs 'evidently with eggs or young' on one Cairngorm hill, and in 1892 William Evans estimated ten pairs 'in a very limited area of the Cairngorms'. One trip, Evans recorded, 'consisted of no less than 14'.

Between 1900 and 1919 dotterels nested on all main hill groups in the Cairngorms, but I have no details of numbers.

From the 1920s onwards we began to learn a little more about these populations. In Table 19 I have recorded all data known about breeding between 1920 and 1970.

1920–9
In central and west Cairngorms dotterels nested in high numbers in 1922, 1924, 1927 and 1929 and certainly in lower numbers in 1926.

1930–9
Records for the east Cairngorms are incomplete. But in the central and western Cairngorms 1933 (nine to ten hens on central Cairngorms) was the best year, and 1934, 1936 and 1939 were the years with high numbers. There were fewer dotterels in 1931–2, 1935 and 1938.

1940–9
In 1940 at least fifteen pairs nested on the central and west Cairngorms.

In 1941 fewer pairs nested on the two western and central massifs, but we found four nests on lower western spurs and summits.

In 1942 breeding populations on west and central hills had slumped to half those in 1940. In 1941 and 1942, moreover, there were more hens than cocks on both western hill systems.

I have few records for 1943 and 1944. 1945 and 1946 were not years of peak populations. In 1947, a year of late thaw and heavy snowcover, six to seven pairs nested rather late in the central Cairngorms. 1948 was not a year of high numbers. We failed to locate pairs on four regular nesting-ridges on the central Cairngorms.

In the warm dry summer of 1949, dotterels probably nested in greater numbers and at a higher density in the central Cairngorms than in any of the first fifty years of the twentieth century. On ridges east of Lairig Ghru we found nine nests and two broods.

1950–9
In 1950, a latish breeding season, nesting groups contained roughly half the pairs recorded in 1949. On the other hand, Hobson recorded more pairs breeding in the central Grampians than in 1949.

In 1951, a year of exceptionally heavy winter snow, dotterels nested late, but in good numbers, on the central Cairngorms. In June Van den Bos saw fifteen on the central massif, and in late June six or seven pairs were still courting. In the east Cairngorms dotterels also nested late and in fair numbers. Our survey of the western tops was incomplete, but three pairs nested on the lower spurs.

In 1952 dotterels apparently did better on west than on central and eastern Cairngorms. On the central massif we located four to five pairs in a latish breeding season. 1953 and 1954, bumper breeding years in the central Grampians likewise produced high numbers on west and central Cairngorms, but there are no records of high numbers on eastern bens. In 1954 Adam Watson watched a party of forty dotterels flying over the 'Mar side of the mountains' – a heavy snowstorm had presumably caused several groups to gather.

Between 1955 and 1959, to the best of my knowledge, dotterels did not nest at high density in the Cairngorms. The entire annual population possibly did not exceed fifteen pairs.

Heavy snowfalls in mid-May, 1958, delayed nesting on the central Cairngorms, but more dotterels nested there than during the previous three years. We located five to six pairs. Dotterels also nested in good numbers on western tops, but were slightly under strength on lower western hills.

Incomplete records for 1959 suggest that breeding strength probably continued at the 1958 level.

The 1950s were thus a good decade in the Cairngorms, with high numbers in 1953, 1954 and rather fewer pairs breeding in 1950, 1956 and 1957. The total Cairngorm breeding population possibly fluctuated between fifteen and twenty-five pairs, with a decade mean of about twenty pairs. This is, however, only an approximation.

1960–9

In 1960 Cairngorms groups carried high numbers. About ten pairs nested on one east Cairngorm massif; five to seven pairs in the central Cairngorms; and at least four pairs on a part of the west Cairngorms.

1961 was an even better dotterel year. At least nine, and possibly eleven, pairs nested on the central Cairngorms, and as many as ten pairs on one eastern massif. In the west there were also high numbers. In 1961 about thirty to thirty-five pairs nested.

In 1962 numbers dropped. The central Cairngorms still carried six to seven nesting pairs, but in the east and west the breeding groups were down. Numbers continued to fall in 1963 with fewer nesting pairs on all hills investigated; but in 1964 it was better – four to five pairs nesting on the central hills. Nevertheless, the breeding population there was less than half that of 1961. 1965 was another moderate year.

1966 was certainly not a year of peak numbers. 1967 was a good year, with roughly eight nesting pairs in the central and about ten pairs in the east Cairngorms (A. Watson, *pers. comm.*). I have less data for the west Cairngorms. In 1967, however, probably at least twenty-five pairs bred on the three hill groups.

In May and June, 1968, an abnormal amount of winter snow overlaid the high Cairngorms massifs. In central and eastern Cairngorms there were more hens than cocks. In the central Cairngorms, possibly four pairs nested and on one east Cairngorms massif about seven pairs, but breeding success on both groups was extremely low (0.2 : 1). I have no data for other eastern hills and little from the west Cairngorms. This was, however, clearly a year of poor breeding with a total nesting population of under twenty pairs.

In 1969 dotterels were in high numbers. At least six pairs nested in the central Cairngorms and five pairs on one part of an east Cairngorm hill. (A. Watson, *pers. comm.*)

In the 1960s, probably due to cooling climate, dotterels thus had their best decade of this century. 1960–1 and 1967 and 1969 were years of particularly high numbers. In all Cairngorm groups 1970 was possibly the best year in the century. In the central hills twenty-three dotterels were located, twelve cocks and eleven hens – roughly the same population as in 1949 and 1961. On one big eastern hill there were twenty-five birds – all different – twelve cocks and thirteen hens – one of which had laid her first egg by 13th May.

A bird watcher also reported seeing sixteen dotterels on a western top.

1971 was another wonderful year with a breeding success of 0.4:1 in the central Cairngorms, but only 0.3:1 in the east Cairngorms.

1900–71

In the last fifty years high numbers of dotterel have nested in the Cairngorms in at least eighteen years. In 1949, 1960 and 1961, thirty pairs or more nested, and in 1970 and 1971 not less than fifty. In less productive years like 1926, 1931, 1942, 1955 and 1968, breeding numbers probably fell to half those in peak years. Since 1945, however, twenty to twenty-five pairs is probably a conservative estimate of the average annual Cairngorms breeding strength.

Central Grampians

In the early 1870s MacGregor, a Drumochter gamekeeper, spoke of small trips of six to twelve dotterels 'indicating about the number of nests which occurred within a radius of a few miles of that watershed'. (H.-Brown, 1906.)

In the 1880s there are no reliable records, but in some of the 1890s dotterels nested in high numbers.

1900–19

Records for these two decades are incomplete, but in 1908 E. P. Chance and J. M. Goodall found three nests in three days spent on west Drumochter hills.

1910 was a bumper year. Edgar Chance and Gamekeeper Kennedy found three nests on a western massif. Then, on 26th to 27th June, Gerald Tomkinson located seven broods there. In 1910, therefore, at least ten pairs nested west of Drumochter. Chance also found a nest on an eastern spur. In 1910, no less than eleven pairs then nested on a small area of the barrens. In that memorable year how many pairs really did nest there? Including Gaick Forest, the total breeding force probably consisted of at least twenty pairs.

In 1911 numbers were down. Gerald Tomkinson and Bertie Lings found only two nests on the same ground.

I have few other records for this decade. Some time between 1912 and 1916 (no year specified) Clive Meares found four nests on the east Drumochter bumps.

In 1915 E. S. Steward found five nests, two east of Drumochter and three on hills west of the pass.

There are many gaps in these records, but 1910 was clearly a bumper season. High numbers also probably nested in 1908 and 1915. 1911 was clearly less productive.

1920–9

In the 1920s many dotterels continued to nest. In 1922 Norman Gilroy found seven nests on eastern spurs and in 1924 Percy Smyth and J. R. Pelham-Burn recorded ten nests on the western and eastern hills.

1925 and 1926 were years of lower numbers, but in 1927 there were a good many dotterels. Chance found five nests, three on a western massif and two in the east.

I have no records for 1928. On 28th May, 1929, however, Edgar Chance and Douglas Meares found five nests on one western ridge and Sir Maurice Denny found four nests on the eastern group – a total of nine nests and much suitable ground unworked.

In the 1920s dotterels nested in high numbers in 1922, 1924, probably 1927, and 1929. 1926, on the contrary, was a poor year.

1930–9

In the 1930s there are no records of peak populations. 1931, 1933 and 1938 were possibly the best years.

1940–9

In 1940 Dr George Franklin watched a trip of five hens and found three nests on eastern hills. In 1942 numbers were low: Franklin found only one nest. In 1943 George Yeates, who worked east of the pass, did not report high numbers.

In 1945 Franklin worked the hills on both sides of the pass. In the east he found only one nest; in the west he found no nests, but saw five grass widows.

In 1946 and 1949 Professor William Hobson reported few pairs. In 1946 he found one nest and in 1949 three nests.

1950–9

In 1950 Hobson located at least six, and possibly seven, pairs; three to four pairs in the east and three in the west.

In 1951 dotterels nested exceptionally late in the central Cairngorms, but at usual dates in central Grampians. At least six pairs nested on west and five on east Drumochter Hills. Fewer pairs nested in 1952, but not less than four pairs in the west, and three or four pairs in the east.

In the central Grampians 1953 was probably the greatest dotterel year of the century. Not less than eleven pairs nested on west Drumochter Hills and at least eight pairs on eastern spurs. There were also non-breeding cocks, hens, pairs, trios and small groups. At least two to three pairs also nested in Gaick Forest. From nowhere in the world have I obtained records of a higher breeding density than that proved in the central Grampians in 1953, where I estimate that between twenty and twenty-five pairs nested.

Fewer pairs nested in 1954, but west Drumochter carried at least six to seven pairs and east Drumochter not less than seven pairs. I have no records for Gaick Forest. In this second consecutive year of peak numbers at least thirteen to fourteen pairs thus nested. 1910 is thus the only year where breeding density can be compared with that in 1953-4 (see Figs. 7 and 8).

There were fewer pairs in 1955, but at least three pairs nested west of the pass and not less than five pairs in the east. There are fewer precise records for 1956-7, but at least two pairs nested on west Drumochter and two pairs on east Drumochter Hills. 1958 was a good year – three nests in the west and at least four pairs in the east. In 1959 there were fewer breeding pairs.

In the 1950s numbers fluctuated, but the decade norm was clearly higher than that for the 1930s and 1940s. Indeed I know of no decade with a higher mean.

1960–9

I have fewer records for the 1960s. In 1960, Colin Murdoch watched a trip of eight on an eastern spur. In 1961 Ratcliffe found a nest, saw a pair, and watched a dotterel running – all these on little-worked ground in the west. He also found a brood and saw several dotterels on eastern spurs. In 1960-1 the central Grampians probably did not carry the bumper populations recorded in the Cairngorms, but breeding numbers were high.

I have inadequate data for 1962, but in 1963 and 1964 dotterels nested in fair numbers east of Drumochter. In 1965 and 1966 the breeding population was also fairly high in the west, but lower to the east. In 1967 there were at least five pairs on one central Grampian massif. In 1968 few pairs nested on eastern groups and in 1969 Tewnion and Ratcliffe found no pairs on the western hills where dotterels bred at such high density in 1953-4. In 1970 J. Armitage reported rather low populations on both sides of Drumochter. After much searching he found three nests.

In the 1960s, therefore, probably fewer pairs nested in the central Grampians than in the 1950s, but the groups were larger than in the 1930s and 1940s.

Summary

Between 1900 and 1969 the incontinuity of my records prevents a balanced assessment.

In 1920-9, however, numbers were generally high, but between 1930 and 1949 I am aware of no peak. A phase of warmer climate had possibly changed the dotterel's pattern of distribution.

In the 1950s dotterels probably nested in higher numbers than in any decade in this century. The mean for the 1960s was certainly less, but 1967, and possibly 1961 and 1968 on the west Drumochter Hills were years of plenty.

I give ten to fifteen pairs as the annual average breeding population at present. But I realise that this is little over half that recorded in peak or bumper summers.

West Grampians

There are insufficient records for this wild hill country which includes the Ben Alder group. In 1953, at least five pairs nested on two hills, and in 1968 Ivan Hills located three pairs and found two nests on one plateau. I have records of dotterels in summer on at least three other hills and therefore estimate eight to ten pairs as the annual norm.

East Grampians

In 1908, 1911 and in other years Blackwood found dotterels nesting in good numbers, but I have no detailed estimate.

In the 1950s and 1960s Adam Watson, Derek Ratcliffe, G. Sutherland, Sandy Tewnion and others have watched dotterels. In 1963 Ratcliffe found four nests and estimated five pairs on one hill group. In 1964 there were fewer pairs: Ratcliffe found only one nest. In 1965 there were probably four pairs; Ratcliffe found three nests. 1966 produced four to five pairs, but in 1967 and 1968 there were only three pairs. In 1969 there were four pairs and in 1970 fifteen birds – seven cocks and eight hens (Watson). Between 1963 and 1970, therefore, the average annual population was at least four pairs. This hill group is, however, only a small part of this huge country. In July, 1968, M. J. P. Gregory found two dotterels with broods on tops in Balmoral Forest, in 1969 Sandy Tewnion and Adam Watson reported four nests, and in 1970 eight to ten pairs on these tablelands.

Dotterels also nest above Glen Clova and in the Forest of Atholl.

I am settling for an annual average of ten to fifteen pairs, but in 1970 only two of these groups carried sixteen to eighteen, and probably more, pairs.

South and west Perth

Up to the early 1870s dotterels evidently nested in good numbers in south and west Perth.

In the late 1860s Booth's exploits suggest that fairly high numbers then bred in Glen Lyon. In 1866 he came upon three small trips of 'grass widows' 'consisting each of 3 or 4 individuals'.

Booth also wrote: 'A few days before the 12th of August 1867, six or seven couple were bagged by two guns, while hare and plover shooting in west of Perthshire; and with one or two exceptions a quantity of down still showed round their heads.'

In the late nineteenth century numbers dropped until a cooler climatic phase induced dotterels to nest in south and west Perth (Harvie-Brown, 1906).

In this century there are only sporadic records. We cannot yet count south and west Perth as having significant populations.

Southern Uplands, S. Scotland

In the last fifty years only a few nests have been recorded at long intervals.

ENGLAND

Up to about 1860, fifty to seventy-five pairs possibly nested in England in good years. By 1900 only a handful bred regularly, and from 1927 onwards dotterels have only nested sporadically on English fells (Ratcliffe).

Between 1900 and 1925, however, there were several promising but abortive revivals – seven pairs in 1911, three to four pairs in 1912, at least four pairs in 1917, and at least eight to nine pairs dispersed on four different fells in 1921. None led to resurrection.

Between 1956 and 1971 Lakeland naturalists have found nine nests. One fell held a nest in 1968 and two nests in 1969 and 1970. Rather naïvely, perhaps, I give one pair for England in my estimate.

WALES

1969 produced the first acceptable breeding record.

Summary

In most years the 'breakdown' is possibly roughly as shown in the table below.

NUMBERS OF BREEDING PAIRS (1945-69)

Location	Max. No. of breeding pairs recorded in any one year	Estimated No. of breeding pairs in most years
Sutherland	1	0
Ross	3-4	6-8
Inverness (North of Great Glen)	1	3-4
Monadhliath	2-3	6
Cairngorms, East	15	8-10
Central	9-11	5-7
West	10+	7-8
Grampians, East	16-18+	10-15
Central	22-23	10-15
West	5-6	8-10
Perth, South and West	1-2	-
Southern Uplands	1	-
England	2	1
Wales	1	-
	89-98	64-84

In many years, therefore, about seventy ($\pm$ten) pairs probably bred somewhere in Britain. In years of exceptionally high numbers like 1953-4, or 1961 and 1971, however, over a hundred pairs probably nested. In some of the 1930s, on the contrary, the total breeding population was possibly less than fifty pairs. I hope that this rough estimate may be a basis for more exact measurements in the future.

Population Rhythms and Fluctuations

Table 19 records years when dotterels nested in high or low numbers in one or more regions. Is it possible to distinguish any population rhythms?

Breeding records for 1910–11 require comment. In 1910 at least twenty pairs possibly nested on the Drumochter Hills and at least two pairs on an English fell. In 1911 there were apparently fewer pairs in Drumochter, but in the east Grampians there were high numbers. The Cairngorms also apparently carried large groups. In England, 1911 was the best year in the decade – seven pairs on four fells.

Was there some link between unusually high numbers in England and in the Cairngorms and Grampians? Was there an influx of pioneering pairs? Did colonists from abroad breed in England, but fail to establish stable populations? If there actually were pioneers in 1911 what was their origin? It is unlikely that they were English-bred. In years of abnormally high numbers, for example, do some Scottish dotterels – possibly birds in their first summer – 'hive off' while passing over English fells? But if so, why do they now so seldom settle in England? Is an irruption of 'aliens' during a changing or oscillating climatic phase a more likely explanation? (N.-Thompson, 1966).

The 1920s also pose problems. In 1921 at least eight to nine pairs summered on four different English fells. But these pioneers also failed to form stable breeding-groups. By 1927 dotterels had ceased to nest regularly in England, and now bred only sporadically in the Southern Uplands and in south and west Perth.

In the Cairngorms and Grampians, on the contrary, breeding numbers were often high in the 1920s, particularly in 1922, 1927 and 1929.

In warmer phases of climate, dotterel thus apparently retreat from marginal habitats on distributional fringes.

The 1930s were an uneven decade, with occasional or sporadic sightings or breeding in England, the Southern Uplands, West Perth and Ross, and no bumper groups in the central Grampians. In the Cairngorms, however, 1933 and 1934, possibly 1936 and 1939, were all years of high numbers.

The 1930s were the climax of a brief and possibly temporary improvement in the climate. In these warmer 1930s, when boreal species became scarcer in Iceland and Lapland, dotterels apparently retreated to the higher tops of the Cairngorms. There were years in the 1930s when the entire British breeding-population was probably fifty pairs or less.

In the 1940s there were several notable years. In the warm summer of 1940 dotterels nested well and in high numbers on the central and west Cairngorms, but in only fair numbers on hills above Drumochter. 1941 was another fairly good breeding year in the Cairngorms, but there were surplus non-breeding hens. In the early 1940s one or two pairs nested on Lakeland fells and on Ben Wyvis in Easter Ross.

In the Cairngorms 1949 was one of the most productive dotterel years in the century, but few pairs nested in the central Grampians. At the end of the 1940s dotterels were apparently nesting on more hills than in the 1920s and 1930s, but, apart from the Cairngorms, I have no definite records of really high nesting numbers anywhere. Nevertheless, this cooler decade encouraged pioneers to settle on lower or more westerly hills.

In the 1950s there were several wonderful 'dotterel years'. In 1951 high numbers nested in the Cairngorms and on central and possibly east Grampian hills. On the high massif of the central Cairngorms, however, dotterels nested exceptionally late owing to an extremely deep cover of winter snow. In Drumochter in 1953 dotterels nested at the highest breeding-density ever recorded in Britain. Many pairs also probably bred in the west Cairngorms.

1954 was the second consecutive 'dotterel year'. There were high breeding populations in the Cairngorms and central Grampians, possibly two pairs on a hill in Wester Ross, and a dotterel was seen flying over high ground on the Ross-Inverness march, and two were recorded on different English fells.

In 1958 there were high numbers on the Drumochter Hills, but only fair groups on the Cairngorms, where, on the high massif, heavy snow-cover continued until late June.

In the 1950s dotterels thus began to show up on some slightly lower hills in England and Scotland. In 1956 and 1959 they nested on Lakeland fells as well as appearing on two other hills in 1954. In Ross at least three pairs nested on three different hills in 1956 and on two hills in 1959. They also nested in Ross in 1950 and 1954. Dotterels were also seen in summer on high hills on the Ross-Inverness marches in 1954, 1957 and 1958.

The colder 1960s clearly continued to favour the spread and increase of dotterels in Britain.

In 1960, for example, at least one pair nested in England and many on the Cairngorms – probably ten to fifteen pairs on the east Cairngorms alone. Many pairs also bred in the central Grampians. 1961 was a tremendous year on all Cairngorm groups, but was possibly a little less favourable in the Drumochter. 1963, on the contrary, only a fair year in the Cairngorms, was excellent in the east Grampians. In 1967

stray or pioneering pairs nested in Sutherland and on the Southern
Uplands. Many also nested on Drumochter Hills, on east and central
Cairngorms, and in the east Grampians. 1968, a year of exceptionally
late-lying snow in the central Cairngorms, produced a far less successful
breeding season. On both central and eastern Cairngorms there was a
greater number of females than males. Numbers were also low on east
Drumochter Hills (A. Tewnion), but at least five pairs were recorded on
one central Grampian highland. Dotterels also bred well in the west
Grampians, where at least three pairs nested on one massif (I. Hills).
1968 was also remarkable for the return of dotterels to Glen Lyon Hills,
where at least one, and possibly two, pairs nested (M. J. P. Gregory
and D. Merrie). This possibly suggests a halted migration or return
movement by dotterels during a particularly cold May (*cf.* Harvie-
Brown, 1906). Of four groups of hills investigated in Ross, dotterels
were only recorded on two. But one of these groups apparently carried
fewer pairs than in 1956 and 1959 (Ratcliffe). The pair which successfully nested in Sutherland in 1967 apparently did not return in 1968
and at least one other Sutherland haunt was unoccupied.

1969 was a good dotterel year in the Cairngorms, east Grampians and
possibly in England, but less fruitful in the central Grampians.

In 1970 there were peak populations in all Cairngorm groups, in
east Grampians, and possibly good numbers in central Grampians. A
dotterel was recorded on a 'new' hill in Inverness, north of Great Glen,
a cock obviously with chicks on one Ross hill and seven dotterels on or
over a second hill, and there were sightings on at least three English
fells, on one of which two pairs nested.

How are we to explain these rhythms?

1. Population explosions in Arctic or boreal habitats?

2. Density-dependent factors affecting British breeding populations?

3. Long-term and/or short-term climatic changes or oscillations
affecting biotopes in Britain and in Scandinavia?

Here are the facts.

From the 1940s onwards, dotterels have started to pioneer new or re-
colonise old biotopes in Europe. All this expansion and resettlement has
taken place during twenty years of cooler climate. Surely, then, climatic
changes or oscillations, rather than density-dependent factors, are more
likely to have led to these remarkable spreads and fluctuations.

Conclusions

Has the dotterel's breeding strength in Britain radically altered during the last hundred years?

For England the answer is unequivocal. In the 1800s as many pairs probably nested in England as are now breeding in most years in Scotland. In Lakeland the Mossfool is now a marginal species, nesting irregularly most of the last forty years and only now showing signs of firmer establishment. The Great Slaughter and the Acquisitive Society of the nineteenth and early twentieth centuries, however, were probably not the only causes of the decline. The Little Ice Age of A.D. 1500–1900 gradually gave way to some warmer decades in the first half of the twentieth century. Sensitive to subtle climatic changes, the dotterels possibly shied away from fringe habitats in Lakeland and on some lower Scottish whalebacks. Pioneer pairs also failed to establish permanent colonies and nesting groups on marginal habitats died out. In good years in the Cairngorms and Grampians, however, dotterels probably continued to nest in equally large numbers and at roughly the same density. In the 1950s and 1960s a few pairs have also started to settle on apparently less favourable outlying hills, north, south, east and west of their central Highland heartlands.

If we compare the dotterel's recent expansion against a background of old records, is it unreasonable to speculate that in a cooler climate the Scottish hills once carried at least a hundred and fifty breeding pairs? Apart from man's insensitivity and interference is there anything likely to prevent this again happening in the future? Pioneering dotterel could find almost as many open and welcoming spaces on English fells and Scottish hills as they now seem to find on the new Dutch polders.

And, in the past, I do not believe that all members of those great nineteenth-century spring flocks were inevitably fated to be a non-breeding surplus or mere grist to the mills of the poulterer and tackle-dealer.

CHAPTER 14

Dispersion, Surplus and Breeding Success

THE vast tracts of suitable terrain in the Cairngorms favour dispersion. Dotterels consequently seldom nest at exceptionally high density even in years of high numbers and extensive snowcover. In 1949 and 1961 there were about five pairs to the square mile of dotterel ground in the central Cairngorms. In both years the nests were never less than about three hundred yards apart. In 1936, a year of fairly high numbers but heavy snowcover, three pairs nested in an equilateral triangle, the sides of which were about two hundred and fifty yards long.

On the high tops of the west Cairngorms pairs are similarly dispersed, but in years of low numbers two pairs nest a mile or more apart. A small cluster of two or three pairs, however, often nest on one whaleback, but their nests are often separated by over five hundred yards. Other ridges slightly lower down usually carry only a single pair, except in summers with peak numbers or particularly heavy snow. Two pairs then sometimes nest on the same ridge, but are usually over three hundred yards apart.

Dotterels have the same pattern of dispersal on the lower bumps and ridges of the west Cairngorms. In 1941, for example, we found four nests in a line of two to two and a half miles of stony and mossy tops. Two nests were less than a hundred yards apart, but on one a hen was sitting on infertile eggs.

East Cairngorm groups scatter in the same way. In 1970 about twelve pairs nested on a broadback about two and a half miles long. In the east Grampians there is a similar scattering, but possibly a higher population-density.

On the central Grampians, possibly because there is less suitable ground, dotterels tend to nest in small clusters of up to three to five pairs. These groups usually nest on slopes below the crests of hills where they had previously courted and to which they later take their broods. On one hill-group in 1953 eleven pairs nested on a big whaleback. On one of the hills there were five nests in under a hundred acres. In 1953 three pairs also nested in an area of about a hundred and fifty acres on another central Grampian and in 1954 breeding density was a little less – six to seven pairs breeding on the whaleback where eleven had nested in 1953. On another hill-system five pairs also nested within a square mile in 1954.

Dotterels apparently formed small nesting-groups on English fells, but seldom at high density. In 1888 John Watson remarked that '5 or 6 pairs bred at no great distance from one another' – a rather vague and indefinite statement. In 1911, however, J. F. Peters found two nests about a hundred yards apart, and in 1969 there were also two nests within about two hundred and fifty yards. Breeding density in England thus occasionally equals that on the Cairngorms in good years.

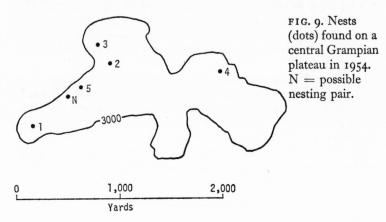

FIG. 9. Nests (dots) found on a central Grampian plateau in 1954. N = possible nesting pair.

I have insufficient records to assess breeding-density on other Scottish hills. In 1956, however, two nests in Ross were half a mile apart, but on different ridges (Ratcliffe). On Scottish hills and English fells, however, a few isolated pairs nest a mile or more from any neighbours. I have fewer records of population-density in Scandinavia and Finland or for the Soviet Union.

On Hardanger Vidda, Norway, Hugh Blair tells me that dotterels were as plentiful as golden plovers.

1. A party of four walking across one fell found six nests.

2. Helthke and his friends found three nests on one ridge.

3. I once lay up beside a little tarn in the early morning and found that dotterels flighted in to feed. They were arriving in parties of from four to eight birds, and I must have seen over thirty dotterels. Of course, with so much suitable ground, dotterels may well be more scattered than in Scotland.

In Finnmark, Professor Hobson found that dotterels nested in higher concentrations than in the Cairngorms or Grampians. 'On the summit of a hill only about 50 feet above the road we found 3 nests within 150 acres.' There the dotterels appeared to be almost colonial.

In Sweden population-density varies locally and seasonally. In good years, however, dotterels certainly nest on Lapland fells in high numbers.

In Finnish Lapland, Palmgren informs me that there are possibly one to two pairs to one square mile of suitable fell, but adds that Meri-kallio's estimate of 4,000 pairs was 'calculated on the basis of a single pair recorded on one linear transect!'

Dr O. Hildén writes: 'The snowcover might be an important factor in spacing out pairs and causing differences in population-density from one year to another. In 1966 I found six nests in an area of 0.9 kms. and almost certainly overlooked others. Two nests were about 200 metres apart.' This is very similar density to that sometimes recorded in the Cairngorms.

I have no precise records about breeding-density in U.S.S.R. In some tundra habitats, however, dotterels nest at high density. Maud D. Haviland (1917) found two nests and heard a third dotterel calling, within half a mile. H. Trevor-Battye also shot two dotterels from nests only two hundred yards apart.

But I have yet to read of higher concentrations than those in the central Grampians in 1953. The polders of Holland may yet amaze us

The non-breeding surplus

Do high density dotterel populations sometimes carry a floating non-breeding surplus?

In the Cairngorms I have never known a breeding pair to occupy every previously occupied territory. There were thus always fewer pairs than potential territories. In some years, therefore, dotterels presumably bred at densities below that which food resources could maintain, although heavy snowcover is sometimes a limiting factor. Non-breeding dotterels, however, are difficult to identify because hens gather in small bands while the cocks are sitting. Mated as well as unmated cocks also sometimes join these groups. Individual recognition is thus less certain than in species with orthodox territorial procedure. Nevertheless, we have watched hens leave trips of grass widows and then make advertising flights over great tracts of hill. Two late pairs probably formed up in this way. We found their nests on previously unoccupied ridges.

Do Scottish groups now breed at too low densities to spark-off self-regulating processes? In summers with heavy snowcover dotterels sometimes breed closer together than in snow-free years. Restriction of suitable nest-habitat and feeding-grounds may thus inhibit less

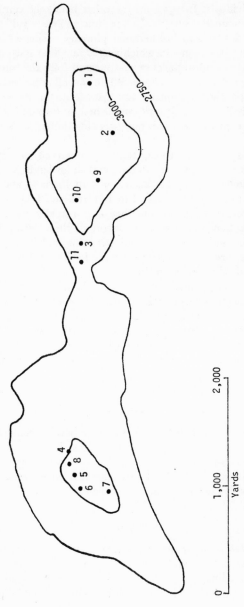

FIG. 10. Maximum nesting density recorded in Scotland. The nests are shown in the order that they were found. (Central Grampians 1953.)

dominant individuals. In 1953, on the other hand, the central Grampian groups contained a surplus. In late June and early July we located non-breeding dotterels in full summer plumage – single birds, pairs, trios and small trips – some of which nested late after most other pairs had broods; others did not nest or disappeared from the tops.

Non-breeding dotterels also appeared on several apparently marginal nesting hills on the Grampians and Monadhliath. Were these the rejects of the high density groups established on favoured whalebacks? And does this explain occasional non-breeding dotterels on low hills in Glen Esk?

The great spring gatherings, formerly observed on the Solway Marshes, Yorkshire Wolds, and in the Scottish lowlands, never seemed in line with contemporary nesting records in England and Scotland. More pairs admittedly then nested in England and bred on a greater number of Scottish hills, but did they ever actually breed at higher densities in Scotland? Did many survivors of these spring gatherings actually form a non-breeding surplus? Does this explain why the Great Slaughter never prevented dotterels nesting in their Scottish heartlands in approximately the same number as now?

In years of peak populations, therefore, do self-regulating mechanisms create a floating surplus? Is a pool of potentially-breeding birds really an index of a strong and healthy population? Do these subtle pressures determine how many pairs will breed in an area in a given year or do they only become decisive when numbers exceed resources? Is a non-breeding floating surplus entirely composed of birds that have never previously bred? Here are a few facts.

Greenshank

Greenshank populations in Inverness and Sutherland sometimes carry a small surplus containing potential breeders. In 1934 and 1937 in Inverness and in 1964 and 1969 in Sutherland, for example, four cock greenshanks found, re-mated and successfully nested with new hens after predators had destroyed their first mates.

Dunlin

In Alaska Holmes (1969) shot a number of territory-holding dunlins and then found that the survivors quickly found new mates among the surplus non-breeding groups. 'Some members of the population were being prevented from settling by already established individuals.'

Golden eagle

In the 1930s golden eagles in Strathspey likewise carried a floating reserve which enabled survivors of broken pairs to find new mates. Between 1936 and 1939 there was a different hen in every year in one territory. In 1937–9 a golden eagle population in Argyll also maintained a surplus capable of replacing losses.

Passerines

A non-breeding surplus is a characteristic of groups or populations of raven, crested tit, siskin and Scottish parrot crossbill. Ravens can replace losses during the breeding season in both high and low density populations (Personal observations and D. A. Ratcliffe, 1962). Non-breeding groups in the cardueline finches are likewise regular components of breeding populations at almost every level. Strong mechanisms of dispersion possibly compel crested tits to breed at densities well below that which their food supply could maintain during favourable years. A hierarchy in the winter flocks possibly evolves a breeding élite and a floating surplus. Territorial behaviour then disperses the breeding groups. After severe winter losses the floating surplus restores stability. By 1949, for example, crested tits in Rothiemurchus and Abernethy Forests were breeding in almost as high numbers as in years before the catastrophic slump in the severe winter of 1946–7.

Membership of the non-breeding surplus

Wynne-Edwards (1962) argues that self-regulating mechanisms are likely to affect breeding numbers in any particular location in any one year, because a habitat probably has an annually varying carrying capacity. 'Usually this number is somewhat less than the total population of potential breeders, so that there is a surplus of non-breeders which cannot win the property and status which entitles them to nest.' The surplus thus consists of individuals which have failed to make the grade in the breeding hierarchy. 'The majority of them are probably junior in age; those that have bred before, we know, often possess a virtually inalienable right to return to and occupy their former sites and territories. . . . It is likely at times that not all members of the excluded non-breeding surplus are mainly individuals that have never yet bred.'

A hen dotterel probably sometimes fails to lay eggs in consecutive summers. Two hens, which I traced for three and five years respectively, had one-year gaps in their breeding sequences.

Greenshank

In greenshanks there is stronger evidence. In 1932–4 a hen greenshank annually laid four unique eggs, each zoned by a distinctive green band. In 1935 I watched two hens in this territory fighting fiercely, the cock apparently pairing with the victor, but I failed to find the nest. Yet the brood was reared in the usual wet flow. In 1936 and 1937 I found the nest which now contained the eggs of a new hen. Green Bands did not nest. In 1938 the territory-holding cock attracted two hens which nested about a hundred and fifty to two hundred yards apart. I watched the trio site-selecting. One hen was Green Bands. Presumably she had lost out in 1935 and only regained her status in 1938.

In another territory in 1935, on the contrary, I watched four distinctive eggs hatch out. In 1936 I found that a new hen had laid in the 1935 scoop. Imagine my astonishment later when I discovered that the 1935 hen had laid in another territory half a mile away. Her new mate had two hens whose nests were a hundred and fifty yards apart.

In 1964–5 in Sutherland a hen with unusual behaviour and distinctive eggs laid in the same scrape. In 1966, a year of low numbers, she failed to breed. In 1967 she laid her eggs in a new territory about one mile away. In 1968 she again failed to breed, but in 1969 returned to the original 1964–5 territory. Possibly an old bird, she laid only three eggs in 1965 and 1969. In 1970-72 a new hen held the territory.

These examples of periodic non-breeding are not strictly comparable, but probably do indicate some instability in breeding hierarchies.

Hobby

In 1930 and 1931 I found a hen hobby which laid particularly well-marked and distinctive eggs late in June. I could have picked them up out of a tray containing scores of others. In 1932 – a year of bumper numbers – a new hen nested in this territory. This hen laid eggs with a strong family likeness to those of the 1930-1 hen and she also laid late in June. In the first week of July I found that the first hen had just started to sit about one and a half miles from her old nesting wood. Was this an example of a daughter ousting her mother from the original family territory?

Raven

In 1949–53 Simson (1966) studied the eggs of an expanding raven population in Wales. In 1949 he found 'four uniquely marked eggs, a perfect identification of the hen bird for all her laying life.' In 1950 a hen in this territory was shot while leaving her nest. No one appears to have seen the eggs. In 1951–2 a hen laying an entirely different type of eggs occupied this territory. But in 1953 the original (1949) hen had returned and laid her 'four most unusually marked eggs in the fir tree nest'. In 1950, therefore, another hen had supplanted the bird laying the unusual eggs. After the supplanter had been shot a third hen replaced her in 1951–6, and then in 1953 the original hen regained her previous territory.

Between 1939 and 1953 this population of ravens was breeding at increasingly high density and was apparently carrying a non-breeding surplus. Failure to hold her place in the breeding hierarchy had thus condemned this particular hen to three consecutive non-breeding years. What proportion of this expanding population had the social guillotine actually inhibited?

Conclusions

We have far to go before we begin to understand the subtleties of the dotterel's dispersion mechanisms. I believe, however, that an intensive study of breeding groups in Holland could be rewarding. On 15th May, 1962, for example, a group of forty dotterels were seen together on a field, and on 18th May, 1963, there were two large groups, one of eighty to a hundred, the other of about fifty, congregated on a wheat-field. Large groups and parties have been recorded in May in sub-sequent years. (J. F. Sollie.) In every year after the first colonisation in 1961 a few nests have been found or broods recorded. I am not convinced that all these large groups consist of breeding birds. They remind me of the great nineteenth-century spring gatherings. If so, they possible contain a large non-breeding surplus, which the machinery of dispersion is restraining or inhibiting.

Breeding success in different populations

Table 20 summarises records of breeding success of groups on different hill-systems in terms of ratio between fledged young and adults. What data we have suggest that groups breeding on the Cairngorms (parti-cularly the high massif of the central Cairngorms) are considerably less successful in producing flying young than those on the east Grampian hills. Breeding groups on the east Cairngorms and the lower ridges of west Cairngorms and Drumochter Hills are evidently more successful than those on the high central massif, but markedly less productive than the east Grampian populations. It is possible that young dotterels also grow more rapidly and fly sooner on the east Grampians than in the central Cairngorms. Data are too few to be significant but they do indicate this possibility.

To what extent does the soil with its influence on vegetation and consequently on invertebrates influence breeding success? Dotterels appear to breed less successfully on the purely granite tops of the Central Cairngorms (0.2 : 1) than on the Moine series of gneisses, schists and quartzites in the Central Grampians (0.4 : 1). They apparently thrive best (1.0 : 1) on the plant communities growing on Dalradian schistose east Grampians (Table 21).

I have insufficient comparable records for Scandinavia, but Hildén (*pers. comm.*) emphasises 'great annual variations in breeding success possibly owing to differences in weather conditions during the critical early stages of the chicks' development'. In 1966 breeding success was extremely high. In five nests only one egg was unhatched and there were two or three young in all seven broods found on 11th July.

If east Grampian groups produce a disproportionate number of fledged young, they may build up a floating reserve capable of strengthening less successful groups or colonising empty spaces on other hill-systems. In this context the dotterel's weak *Ortstreue* is clearly an advantage.

Other montane birds in years of high dotterel populations

Table 22 compares ptarmigan breeding numbers in years when there were large dotterel populations on the Cairngorms and west Grampians.

CHAPTER 15

Mossfool's World

In England and Scotland you can usually recognise typical dotterel hills from a great distance; they have broad flat tops or long rounded ridges and saddles, looking almost like enormous stranded whales. But dotterels are scarce and selective birds nesting on comparatively few of these huge whalebacks.

ENGLAND

In Lakeland many sharp and angular fells are unsuitable but topography and type of ground are more important than altitude. Many haunts are littered with stones and covered with numerous small hummocks. Many pairs nest close to stones; a few others on damp peaty ground.

In the last twenty-five years tourists and hill-walkers have overrun many old haunts. The tops of two famous dotterel fells are now like well trodden football pitches and are covered with tins and broken bottles (E. Blezard).

On the English fells dotterels have few nesting neighbours; but golden plovers and dunlins nest on the tops. On a Buttermere fell, Baldwin-Young found a dotterel and golden plover nesting only twenty-five yards apart. In 1959 Ratcliffe also found five pairs of goldies nesting between the main watershed and the top where a pair of dotterels had their nest. In 1960 and 1968 golden plovers again had eggs or chicks on a dotterel spur. Dunlins also nest on these tops.

Skylarks and meadow pipits haunt the high grasslands and a few wheatears breed on block-littered ground. Lapwings occasionally nest up to 2,700 feet on limestone pasture. On the lower ridges are curlews and snipe and on stony streams dippers and grey wagtails. Ring ouzels nest on the 'scars'.

Birds from the valley sometimes visit the fells. Swifts hawk over high plateaux and rooks work the slopes.

Lakeland dotterels have many natural enemies. In 1920 J. F. Peters watched seven ravens, four peregrines and five buzzards in the air at the same time. In England, moreover, ravens are possibly serious predators. Peters and G. W. Temperley found three nests in which ravens had apparently broken and sucked the eggs. Peregrine, raven, kestrel and buzzard all nest on lower scars, but I have no records of their taking adult dotterels.

Dotterels also contend with carrion crows, rooks, jackdaws and black-headed gulls, which regularly work fell tops in June.

Foxes, stoats and sheep dogs also probably occasionally harry them. In 1926 a fell sheep trod on a nest and eggs.

Derek Ratcliffe describes the dotterel's altitudinal range in England, where, 'despite the less alpine character of the nesting ground, the average altitude of the nests is lower than in the Highlands. Only a few hills exceed 3000 feet in Lakeland and most nests were placed at 2500–2950 feet, though there have been a good many at 2300–2400 feet, and occasional nests have been found as low as 2100 feet during the present century.

'Probably the general appearance of the ground is more important in attracting dotterel than the exact nature of the vegetation. It could be that management for sheep, by producing alpine grassland or "grass heath", has created an attractive-looking kind of vegetation for dotterel at an unusually low level in place of shrub and heaths which are less attractive to them.' Old records also suggest that dotterels formerly sometimes nested at under 1,500 feet on quite low fells.

Southern Uplands

In 1967 Donald Watson's nest was in very short grass on a hill shoulder at about 2,250 feet, well below the summit plateau. This haunt was grassy, quite tussocky in places, with some blaeberry and club moss around, but was much less bare than the summit plateau with its close mat of dwarf willow and blaeberry. After the two chicks hatched the cock led them over 1,000 yards to the summit ridge where they probably found more insects on the barer and more stony ground. Once the little cock ran aggressively at a skylark which landed nearby.

Grampians

In the central Grampians dotterels have several different haunts. The absence of peat on many of the hills west of the pass and its presence on the east have helped to create different bird communities. On the eastern spurs there are often large groups of golden plovers and dunlins. In 1955 I found a dotterel and dunlin sitting just five yards apart. In good years there are also many ptarmigan and meadow pipits and a few skylarks. Red grouse nest on grasslands higher up than the lower-nesting ptarmigan groups. A few wheatears nest on eroded stony shoulders. In summer red deer and black-faced sheep feed on the tops and flanks.

East of Drumochter most dotterels nest between 2,700 and 3,000 feet, but a pair occasionally nests in short outcrop heather on or about the 2,600 foot contour.

Across the pass the western broadbacks have more character but fewer birds. Apart from dotterel and ptarmigan many stony ridges are almost birdless. A few dunlins and golden plovers nest on eroded tracts of peat, but these are far fewer than on the long eastern chain. There are meadow pipits in gullies and wheatears on stone-littered screes. On lower heaths red grouse and meadow pipits dominate and cock ring ouzels pipe from bluffs and falls. Red deer and black-faced sheep share the high grazings.

Dotterels nest from about 2,850 to 3,200 feet, just occasionally lower or higher.

These dotterels have few natural enemies. Golden eagles nest on both sides of the pass and merlins on a western hill flank. Peregrine, merlin, golden eagle, hooded crow and fox are all potential predators.

Some dotterel groups in Angus and Aberdeen possibly do best of all British populations. The rich flora on some of these hills probably carries a great wealth of insects. Dunlins and golden plovers nest in high numbers but their nests are more widely spaced than on the east Drumochter tops. Ptarmigan nest quite low down as well as on the summits and high spurs. Their populations fluctuate but in peak years they breed particularly successfully. Skylarks and meadow pipits share the high ridges with dotterel and ptarmigan. Ring ouzels nest in some corries.

Golden eagles and foxes are likely predators. Golden eagles breed at a density of one pair to about 10,000 acres and foxes at about one pair to 3,000 acres. Stoats are scarce and wild cats absent. In July, 1967, however, a peregrine falcon killed a dotterel on one high top. Crows occasionally rob ptarmigans' nests and might rob dotterels'. No ravens work the high tops. In June, 1965, a fox or shepherd's dog apparently raided a nest.

On one massif dotterels nest on a high plateau at above 3,700 feet. On another group their nests are usually found between 3,200 to 3,400 feet and seldom below 3,000 feet. These groups, therefore, nest higher than those on the Drumochter Hills, but usually at lower elevations than on the central Cairngorms.

Cairngorms

In the east Cairngorms the dotterels nest on massive whalebacks and

tops with knobbly tors. Here they have few neighbours, except ptarmigan. But there are a few meadow pipits. In vast corries snow buntings
also sometimes nest.

In the central Cairngorms dotterels often nest on stony ground,
sometimes laying on extensive granite screefields. Pairs also occasionally
nest on open gravelly ground close to melting snow. Each Cairngorm
hill group has slightly different bird communities. On the high tablelands of the central Cairngorms, where there is little or no peat, dotterel
and ptarmigan are almost alone. Wheatears and a few snow buntings
nest in the corries and meadow pipits on falls and gullies. In the west
Cairngorms the dotterel is less lonely. Golden plovers and dunlins
nest in the fringe moss on these rounded hills. A few wheatears nest on
stony falls and meadow pipits in alpine grasslands. A little lower, the
peat-hags of Am Moine Mor usually carry large nesting groups of
dunlins and golden plovers. Meadow pipits and skylarks also nest
there.

A pair of common sandpipers sometimes nests close to Loch nan
Cnapan, and common and black-headed gulls have nested on Loch
Stuirteig. The high west-central Cairngorms are largely birdless. A few
dotterels and ptarmigan share the flats. Wheatears nest in the screes
and meadow pipits in tongues of grass. Perry and Watson have watched
ring ouzels and broods above 4,000 feet. Snow buntings sometimes fly
over the dotterel grounds on their way to nests in great corries below.
In 1941, at about 4,000 feet above sea-level, I also watched a very tame
dunlin with a greyish white nape and large black patches on its belly –
possibly a northern bird from the fells of Norway or Sweden.

East of Lairig ptarmigan also share the higher ridges and tablelands.
Wheatears and meadow pipits do not really belong, but ring ouzels
sometimes nest high up on bluffs. Lack of peat discourages golden
plover and dunlins, but in 1963 a pair of black-faced northern-type
goldies hatched chicks and in 1962 and 1963 a stray pair of dunlins
nested on a ridge at about 3,600 feet.

On the higher and more stony ridges and humpbacks, where the
dotterels nest, wheatears, and sometimes snow buntings, are their
nesting neighbours as well as ptarmigan. I have found dotterel, snow
bunting and ptarmigan all nesting within a radius of 250 yards. In late
summer, meadow pipits also often lead their broods up to the high
grasslands.

In summer hinds graze on grassy ridges and red stags seek snowfields
in the heat. A herd of reindeer of Swedish Lapland origin often graze in
the dotterel country in the central Cairngorms.

When I first watched the hill birds, black-faced sheep seldom reached
the tops, but in post-war years small flocks and stragglers regularly feed

on the central tops. Blue hares are few in dotterel country. But I have sometimes seen one loping over a ridge where a dotterel was sitting.

On the Cairngorms golden eagles are the commonest bird predators. A few pairs have eyries above 3,000 feet. 'About one-fourth of the hunting ground of eagles in this region is in the arctic-alpine zone and out of 17 pairs in the Cairngorm-Braemar area only one does not have arctic-alpine ground in its hunting range.' (L. Brown and A. Watson, 1964.) I have no evidence, however, that eagles often take dotterels.

On the lower slopes of the Speyside Cairngorms, three to four pairs of merlins nest in heavily-grazed remnants of old forest well below 2,000 feet. At least '5 or 6 pairs are known to nest on the Dee side of the Cairngorms, all in glens or in low hills below 2000 feet'. (A. Watson, 1966.) Merlins are possible predators but the occasional wandering kestrels are unlikely to harm the dotterels.

In summer ravens are seldom seen on the tops, but in the central Cairngorms mongrel crows sometimes rob dotterels' nests.

Cock snowy owls sometimes summer on high Cairngorms plateaux where so far they have preyed almost entirely on ptarmigan.

On all Cairngorms dotterel hills the fox is the only important mammal predator. Stoats and wild cats are scarce.

In the Cairngorms the dotterel's altitudinal range varies. On one eastern whaleback most pairs nest between 3,700 and 3,900 feet contours (occasionally as low as 3,100 feet). On a second eastern group nests usually lie between 3,600 and 3,800 feet.

In the central Cairngorms a few pairs regularly nest above 4,000 feet, but most breed lower down. One long ridge, where two pairs sometimes nest is roughly 3,400 feet above sea-level, but most nests are between 3,700 and 3,900 feet. On lower western hills dotterels favour ridges between 3,250 and 3,400 feet, but odd pairs settle between the 3,000 and 3,100 feet contours.

Ross and Cromarty

On wild wind-swept hills of Ross three nests were between 2,800 and 3,000 feet. A trip of dotterel also ran over peat lands about 2,450 feet above the sea. These hills carry small groups of ptarmigan on the high heaths and a few golden plovers and many meadow pipits and some skylarks on the grasslands, and a few wheatears on the screes. It is a country of mixed deer forests and sheep runs.

Golden eagles, possibly peregrine, and fox are likely predators.

Sutherland

On the Sutherland dotterel hill of 1967 ptarmigan sometimes nest in fair numbers. A few pairs of dunlin and golden plover also haunt the high grasslands and peat-hags, as do skylarks and meadow pipits. Wheatears haunt stony places.

This hill belongs to a huge sheep farm and straggling groups graze on the high spurs.

Golden eagles hunt the hill. In 1968 an eagle or peregrine killed a fulmar whose semi-plucked corpse lay on the hillside. In August, 1967, Norman Moore watched a merlin apparently 'raise' two juveniles. The haunt is also close to colonies of great black-backed gulls. Hooded crows and foxes visit the top. In 1967 and 1968 there were also at least one pair of bonxies. No peace for dotterels here!

Norway

On the Dovre Fjell dotterels nest on high-level plateaux, covered with reindeer moss, small tufts of grass, patches of heather and moss-covered stones.

On the Hardanger Vidda Hugh Blair tells me that 'there is often a tongue of dotterel ground right between two patches of willow scrub. So you sometimes found a nest quite close to birds one would hardly associate with dotterels. You can expect to find shore larks, a few Lapland buntings, golden plovers, purple sandpipers, wheatears and ptarmigan nesting on pure dotterel ground. In cold years there might also be a snowy owl and even long-tailed skuas. Where willow and scrub are close together you sometimes find dotterels nesting beside willow grouse, dunlins, meadow pipits and bluethroats.'

Blair had the unique experience of standing on the Vidda watching great snipe 'lekking' in the dusk and a ptarmigan displaying over a ridge where dotterels had a nest.

'If there are any boulders you may also find a pair of fieldfares. Some fjells, however, do not carry such a good showing of birds. The high ground on Dovre Fjell, for example, is markedly less favourable.'

On the dotterel fells north of Varanger Fjord, Blair records a galaxy of bird neighbours – golden plover, purple sandpiper, shore lark, Lapland bunting and Buffon's and arctic skuas, great black-backed gulls and arctic terns. He also once found a Temminck's stint's nest on dotterel ground and watched a pair of little stints preparing a nest-scoop on the same fell. Turnstones also sometimes nest on these stony ridges. Twice Blair also found rough-legged buzzards' nests on the ground in dotterel country.

Here are little pockets of swamp. 'I once saw a brood of newly-hatched wood sandpiper chicks in a tiny marsh – no more than 5 yards square.' In Finnmark, solitary pairs of fieldfares sometimes nest on ledges of rock, but redwings are restricted to the scrub.

Dotterels have other interesting neighbours on the Norwegian fells. Red-necked phalaropes nest up to about 4,000 feet in the arctic-alpine zone, but they generally favour wetter and more marshy terrain.

Yngvar Hagen tells me that polar and red foxes sometimes prey on dotterels. He collected pluckings of dotterels at ten nests or feeding places of merlins, at three of hen harriers, and at two of snowy owls. Rough-legged buzzards also occasionally take dotterels. In his thorough research, Hagen examined 1,453 snowy owls' pluckings, thus proving how seldom they prey on dotterels.

Hugh Blair mentions that gyr falcons, rough-legged buzzards and occasionally short-eared owls hunt the dotterel country. Ravens go egg-hunting, as do arctic and long-tailed skuas and great black-backed gulls; all these birds sometimes pick up dotterels' chicks and eggs. On some fells golden eagles hunt the high ground. Merlins seem to kill more dotterels in Norway than in Scotland, but the hen merlin is more often seen hunting during incubation than in Britain. The hen kills most, but the 'jack' also takes almost anything that flies. 'To a Scottish dotterel-watcher, the hen harrier is probably a surprising predator, but on Hardanger Vidda I used to watch a fine blue male regularly hunting, alternating between scrub and bare open fells, just as the merlins did.'

Arctic fox and ermine also hunt Hardanger Vidda. Wolverines destroy many birds there. These are the chief predators in this wonderful Norse fell country.

In Norway dotterels have a considerable altitudinal variation, but do not nest at quite such high or such low elevations as do snow buntings. In Finnmark they breed on small rounded hills at about 300–1,000 feet, but in south and central Norway they seldom nest below the 3,500 feet contour. In Valdres, for example, dotterels nest between the 4,200 and 4,400 feet contours. Their principal breeding grounds on the Hardanger Vidda and Dovre Fjells are also about the 4,500 feet contour. On the other hand, dotterels have nested below 2,200 feet on Averöya.

SWEDEN

In Sweden dotterels are restricted to open, bare, high fells in the Arctic zone and less commonly to similar plateaux at lower levels (P. O. Swanberg *in litt*). In Lapland dotterels not only nest on high plateaux and tablelands, but also on isolated fells rising above the coniferous forest (Blair).

On the Abisko fells the dotterels apparently treated nesting snow buntings and wheatears as ecological competitors. This was surprising as they often fed close to larger snowfields, whereas the snow buntings generally exploited *Diptera* around smaller snowpatches.

On other Swedish fells, ptarmigan and golden plovers are common nesting-neighbours.

Herds of wandering reindeer graze the high spurs. Rittinghaus suggests that dotterels and their chicks have to make long journeys through the interference or competition of these herds.

Swedish bird watchers record little about predators; but fox and ermine are mammal predators and ravens and long-tailed skuas are probable bird predators.

In Swedish Lapland dotterels nest from 1,800 feet at 68° N to about 4,300 feet. Farther south (62° N) they nest upward from the dwarf birch, *Betula nana*, zone about 2,800 feet. (P. O. Swanberg.)

FINLAND

In Finland the biotopes are always dry, flat fjell heaths above the tree-line, a low and scattered vegetation of lichens and mosses and sparse carpets of low micro-shrubs – *Empetrum*, *Loiseleuria*, *Diapensia*, etc. – connected by bare patches of ground with small flat-lying stones (P. Palmgren and O. Hildén).

In Finnish Lapland dotterels nest from about 2,000 feet upwards: no nests have been found at lower elevations.

U.S.S.R.

In U.S.S.R. dotterels have many different nesting niches, but apparently always avoid lowland swampy tundra covered with dense grass or shrubs. In the tundra they usually nest on comparatively dry hilly places, with stony soil covered with short grass or Iceland moss or lichens. Sushkin (1938) also records dotterels nesting in stretches of alpine meadow with soft boggy soil and rich vegetation.

In the alpine zone of Siberian mountains they occupy similar habitats, nesting on flat hill tops with rocky soils and scanty alpine flora. These habitats alternate with swampy alpine meadows and thickets of dwarf birch, *Betula nana* (A. I. Ivanov, *pers. comm.*).

We know little about the dotterels' neighbours in the tundras of U.S.S.R. Maud D. Haviland (1917) watched dotterels in the tundras around the River Yenesei where wheatears and shore larks were nesting in forest-bogs. Dotterels occasionally missed their niches and then nested in willow-scrub in the kind of places snipe might choose. 'One nest was in a marshy spot and another, although not 60 feet above the swamps, was scooped out in alpine flora in soil as dry and stony as the mountain top it resembled in miniature.' At the end of the summer broods congregate in little trips 'on the great open spaces of the tundra where the Asiatic golden plovers were gathering'.

For many miles round the Yenesei the tundra did not vary in altitude by more than 200 to 300 feet.

The dotterel's altitudinal range in U.S.S.R. varies with latitude. On northern tundras they breed at very low elevations, but in the Baikal range they are not known to nest below about 6,000 feet and in Khangai Mountains, in Mongolia, Elizabeth Kozlova found them nesting between 9,200 and 9,800 feet.

No one has apparently studied predators in Finland and U.S.S.R., but Haviland watched a rough-legged buzzard attack a brood near the Yenesei.

HOLLAND

Dotterels now nest regularly in reclaimed polders in Holland. Originally these polders had attracted many dotterels on their spring passage. Between 1961–68 nests were recorded in sugar-beet and among new potatoes on the north-east polder and in flax, sugar-beet, wheat, peas and potatoes in east Flevoland. All the nests were in summer crops in treeless habitats. Dutch ornithologists already have strong evidence that toxic sprays have killed old and young dotterels in habitats in flax. Dotterels have thus joined lapwings, Kentish plovers, avocets, oyster-catchers and other colonists, and are now running the same risk of toxic contamination.

Will these pioneering dotterels establish permanent breeding groups in competition with predators and more versatile species? For example, Dutch farmers encourage kestrels and owls, and stoats and weasels as predators of voles. The dotterels still continue to breed in good numbers, but hatching success is appreciably lower than that in Scotland and Fenno-Scandia. Between 1961–68 at ten nests only sixteen young hatched out of twenty-nine eggs – a hatching success of 55.1% compared with 91.8% in the sample shown in Table 17. Nevertheless, the dotterel has shown resilience and ability to colonise these new habitats which possibly superficially resemble their prehistoric niches in the steppes.

AUSTRIA

In the Seethal Alps the dotterels recently nested in low alpine pasture rich in lichens. These ridges have embedded rocks and boulders but are less stony than northern biotopes. These dotterels favoured long flat ridges between 6,600 and 7,600 feet.

ITALY

In the Abruzzi dotterels nested at over 9,600 feet on a high plateau.

The Italian dotterels have unusual associates. Rock pipits and wheatears are plentiful and alpine accentor and snow finch, and alpine and common chough nest on the falls and cliffs. Golden eagle, kestrel and hobby hunt the high flats. The fox is, however, a more formidable predator.

POLAND

The 1946 nest in the Riesengebirge was just above the 5,000 feet level and on a fell rich in boreo-alpine insects and covered with tundra-lichen and alpine meadow.

Hooded crows and foxes are the most likely predators.

SPRING PASSAGE HABITATS

On spring passage to their nesting haunts, dotterels sometimes rest and feed on great shingle beaches on the sea coast, but they seldom mix with other waders on the shore.

John Walpole-Bond (1938) recorded that 'the sole alighting points were usually downland in East Sussex, 8 miles from the coast, though they sometimes rested on "fallow", level, and even big expanses of beach'. In other southern counties dotterels also sometimes rest on downlands. On 14th June, 1927, for example, Walpole-Bond watched one running over a flinty field on Salisbury Plain.

The old naturalists describe habitats chosen on spring passage through England – sandy fields in East Anglia, downs in Wiltshire and Berkshire, wolds and saltmarshes in Lincolnshire, Yorkshire and Lancashire, and beside the Solway Firth.

Trips arrived in the second week in April on the Scottish borders where they rested on particular fields and old leas on upland farms or on land newly sown with grain; on hill farms in Angus, Perth and Aberdeen they spent about two to three weeks in April before flying off to the high tops.

In the last fifty years watchers have recorded dotterels on heaths and rabbit warrens in Norfolk and Cambridge; on ploughed fields, usually sown with oats or corn, in Derby, Lincolnshire and Holy Island off

Northumberland. In Lancashire dotterels now usually favour cultivated lowland moss areas – especially rolled oats fields in late April and May.

Dotterels still sometimes touch down on sandy fields and golf courses in Kirkcudbright and Midlothian. Trips also regularly rest on high grasslands in Brecon, Radnor, Glamorgan and Caernarvon.

We know little about the dotterel's preferences on spring passage through other countries. In Ireland most were recorded on mountain tops. In Co. Dublin on 29th April, 1919, however, G. R. Humphreys saw three dotterels in young wheat about two hundred yards from the sea.

In southern Norway heavy snowcover sometimes forces trips to stay on higher pastures and upland farms until the thaw clears their breeding grounds.

Dotterels apparently make a single flight from the Continent to Swedish breeding fells, but in late springs they spend some weeks on upland farms.

On spring passage through U.S.S.R. dotterels rest on barren steppes, or on cultivated fields and meadows. They sometimes visit gravelly tundra, not far from the sea, in Novaya Zemlya.

In spring, large trips now visit reclaimed polders in Holland and in late April a few rest in freshly worked fields in the Belgian lowlands where they are usually found on beetroot fields – almost the only bare fields in spring passage.

Autumn passage habitats

On autumn passage dotterels feed and rest on various habitats. Cultivated fields now seldom attract them; other biotopes probably provide more insects.

In late summer and autumn observers have recorded solitary dotterels or small trips on sewage farms in Durham and Nottinghamshire; on moorland in Derbyshire; on rabbit warrens in Norfolk; on sea shingle in Kent; on short grass saltings in Dorset; and on the tops of English fells.

In Fenno-Scandia little is known about autumn biotopes. From Swedish fells they apparently branch along the great plateaux into Norway and then move to Denmark where large flocks sometimes rest on Jylland heaths. They are seldom recorded on the coastline.

In U.S.S.R. passage birds begin to arrive in the last ten days of July on tundras close to the Yenesei (Tugarinov and Buturlin, 1911). In autumn great flocks fly straight across the steppes, rising at day-break to feed on cultivated fields (Sushkin, 1908).

In late August and early September groups visit beaches close to
Hook of Holland.

On autumn passage a few also sometimes temporarily rest on farms
in Belgium and Germany and exceptionally on high grasslands in
Switzerland.

What risks and hazards do dotterels face on passage and in their
winter quarters? Man and weather are doubtless the most deadly.
Peregrine, merlin and lanner and sparrow hawk, goshawk, harriers and
the larger owls are all probably dotterel-killers. Foxes, stoats, weasels,
rats, and all manner of cats also doubtless enjoy the occasional mossfool.

Winter grounds

In their winter quarters in North Africa and the Mediterranean basin
and beside the Caspian Sea, wintering dotterels are not true shorebirds,
but select special ecological niches.

In Morocco they favour stony steppes, *Artemisia* plains and newly
ploughed fields in the sub-desert rather than richer farmlands in the
north. They haunt semi-desert plains in Tunisia, where Lilford once
found them consorting with sandgrouse, lapwings, golden plovers,
little bustards and cranes. In Libya they sometimes feed near the sea-
shore in company with golden plovers and in north-east Sinai, in
Egypt, and in Palestine and Syria on semi-deserts and marginal grasslands. Dotterels thus always seem to prefer light, stony or sandy soils
and scanty vegetation.

CHAPTER 16

Breeding Habitat of the Dotterel in Britain

by D. A. Ratcliffe, Ph.D., D.Sc., The Nature Conservancy

General features

THE dotterel is one of the few true mountain birds in Britain, where it breeds only within what may be termed the Arctic-alpine or montane zone, on the upper levels of the higher mountains. In worldwide ecological terms it is a summer bird of northern tundra, heaths and sparsely vegetated barrens ('fjaeldmark') – almost a periglacial species, passing the few clement months each year in habitats of appalling winter severity. Ground of this kind is restricted to high mountains in the southerly parts of the species' range, but with distance north occurs at increasingly lower elevations, and extends down virtually to sea level in the high Arctic.

The reconstruction of British vegetational history since the retreat and disappearance of the Quaternary ice-sheets has particular relevance to birds such as the dotterel. The persistence of an Arctic complex of fjaeldmark and tundra south of the ice, and the migration northwards of this habitat in the wake of the receding glaciers and snowfields, is likely to have had a parallel in the associated avifauna, of which this species is a characteristic member. The distribution and abundance of the dotterel in these prehistoric times will never be known; it may well have been widespread, but one has to note that to-day it is a species which to a marked degree fails to occupy all suitable habitats, both within Britain and throughout its total Eurasian range.

The later general historical picture is of an increasing restriction of the Arctic element of vegetation, flora and fauna to high ground on mountains in Britain as the climate became warmer and forest advanced over the lowlands and lower uplands. After the period of maximum restriction, during the Post-glacial Climatic Optimum, a slight and gradual lowering of temperature is presumed to have been matched by a limited re-advance of the Arctic or montane ecosystem to the levels where it occurs to-day.

Two thousand years ago or less, before the impact of man became really severe, a clear altitudinal zonation of vegetation types would have been visible on British mountains, as it is in western Norway to-day. The forests of the lower slopes gave way at around 1,500 to 2,000 feet to a narrower belt of scrub, decreasing in stature with altitude and showing replacement in turn by dwarf shrub heaths in which ling heather, *Calluna vulgaris*, was probably then, as now, the most prevalent species. Above this came the uppermost zone of alpine grasslands, moss and lichen heaths, often forming an incomplete vegetation cover, and subject to extremely harsh climatic conditions, especially high winds and low temperatures.

Human impact has since caused extensive deforestation of the British uplands, especially towards the upper limits of tree growth, and has almost totally destroyed the high level scrub zone. These vegetation types of the sub-montane zone have been replaced by great sweeps of heather moor, the domain of the red grouse and red deer, or by grass-lands where management has been largely for sheep. The vegetation of the montane zone above has been variably modified, depending on the type and intensity of land use. Where management has been mainly for grouse or deer the earlier zonation remains substantially unchanged, though the widespread practice of moor burning has reduced the abundance of the more sensitive dwarf shrubs. On sheep walks, how-ever, the montane dwarf shrub heaths and even the moss heaths have become modified or changed into high level grasslands of completely different character.

The dotterel's summer haunts are the montane zone of dwarf shrub heaths, short grasslands, moss and lichen heaths, of both original and modified types. The bird is sometimes seen on the lower moors, especially at migration time, but the dotterel which stay to breed appear to live largely above 2,000 feet. Within this upper zone the vegetational pattern varies considerably on any one mountain, especially according to length of snow-cover and drainage. Dotterel range widely over the tops in their feeding, but their requirements for nesting habitat are fairly precise and limited to certain kinds of vegetation and terrain.

On a single mountain, length of snowcover tends to increase with altitude, but is greatly influenced by the differences in shelter and exposure from place to place. Snow usually blows off exposed ground, such as ridges and summits, and piles up in sheltered places, especially to leeward. The mountains with the longest-lasting snow beds are those which are both high and of massive form, with extensive plateaux and flat-topped upper spurs, whence large amounts of snow gather before drifting into sheltered places, ranging from minor hollows to large corries; and there building up to a considerable depth through the winter. As the bird favours the massive type of mountain summit, typical dotterel mountains often carry long-lasting snow beds melting out only in late spring or summer, and this feature adds to the Arctic or alpine character of the breeding haunts. However, dotterel nest on windswept ground and avoid the areas of snow accumulation, some of which still carry snow beds when the bird lays its eggs.

There is often a good deal of wet ground on the high plateaux and ridges, with springs, rills, little marshes, and more extensive areas of peat bog or damp grassland and heath. This, too, the dotterel avoids in choosing its nest site, which is usually on dry ground carpeted wholly or in part by mosses and lichens, with a sparse growth of short grasses and other small herbs, mainly of alpine type. Often the nesting grounds are open, sparsely vegetated and stony barrens where much of the soil and detritus is subject to frost-thaw movements (solifluction) which resist the establishment of a continuous plant cover and often produce a patterning of the ground surface.

The mountain systems in the different upland regions of Britain vary considerably in character. Compared with the Scottish Highlands the hills of England, Wales and the Southern Uplands of Scotland have on the whole been more severely modified by the heavy grazing and burning associated with a long-continued sheep management régime. Most of the higher British mountains are formed of acidic rocks which, in combination with the generally heavy rainfall, give a prevalence of poor, infertile soils with low productivity. There are a few localised calcareous mountains which, by contrast, are notable for the richness of their flora. To the dotterel, however, the most important regional differences in the British mountains are in altitude, topography and climate.

It is the central and eastern Highlands of Scotland which show the greatest ecological affinity to the Arctic regions which are the chief summer home of the dotterel. This district contains the greatest concentration of attractive high-lying plateaux, and is also the coldest and most continental mountain district in Britain, with the greatest development of long-lasting snow beds in summer. Farther north in the Highlands the summits are lower and smaller, with fewer late snow beds, and in the west there is a prevalence of sharp peaks and narrow summit ridges which are unattractive to dotterel. With increasing distance south of the Highlands temperatures become progressively warmer, prolonged snowcover diminishes, and there is a dwindling climatic similarity to the Arctic. There are also few mountains exceeding 3,000 feet, and much of the upland country is relatively low-lying.

Dotterel are distributed accordingly. The bulk of the population breeds in the central and eastern Highlands, there are scattered pairs and small groups elsewhere in the Highlands, and perhaps nowadays only sporadic breeders in other regions to the south.

Another regional trend is relevant to the distribution of dotterel, namely, the altitudinal depression of mountain vegetation zones with distance west and north in Britain. Increasing oceanicity towards the west and decreasing warmth towards the north produce a downward extension of the montane zone, so that suitable dotterel habitats occur fully a thousand feet lower in areas such as north-west Sutherland. It is interesting that the lowest breeding record known in Britain during recent years was for one of these oceanic montane areas.

THE SCOTTISH HIGHLANDS

The Cairngorms

The Cairngorms contain by far the largest mass of really high land in Britain, with four summits above 4,000 feet, namely Ben MacDhui (4,296 feet), Braeriach (4,248 feet), Cairn Toul (4,241 feet) and Cairngorm (4,084 feet), and several others just below this elevation, such as Beinn a' Bhuird (3,924 feet), Ben Avon (3,849 feet) and Beinn Bhrotain (3,795 feet). This is an elevated tableland, rising in places to still higher rounded tops and plateaux, but with steep flanks carved deeply by ice into great corries. The prevailing granite gives coarse-textured, porous soils liable to movement and erosion. In the west, Moine Schists give rather more fertile, silty soils.

Although the tree-line has been generally depressed by clearances, and replaced by sub-montane heather moor, *Calluna vulgaris*, there are great areas of pine forest on the lower slopes, and on Creag Fhiachlach in the north-west, a true natural forest limit occurs at 2,100 feet, with a scrubby growth of pines and juniper passing into the montane heather moor above.

The former zone of montane willow scrub is now represented only by fragmentary growths of whortle willow, *Salix myrsinites*, and downy willow, *S. lapponum*, on some higher cliffs and rocky slopes. Some plants common in the Arctic and alpine regions, such as mountain avens, *Dryas octopetala*, purple saxifrage, *Saxifraga oppositifolia*, net-leaved willow, *Salix reticulata*, holly fern, *Polystichum lonchitis*, and alpine mouse-ear, *Cerastium alpinum*, are in the Cairngorms confined to a few elevated cliffs with calcareous rock, mainly Moine Schists, or crushed zones in granite. In general, in the British mountains, the soils of the high watersheds where dotterel breed are leached and acidic so that lime-loving plants tend to be scarce or absent, and the vegetation is composed mainly of calcifuge or indifferent species.

Within the zone of heather moor, dominance of blaeberry *Vaccinium myrtillus*, and/or northern crowberry, *Empetrum hermaphroditum*, indicates prolonged snow cover. With increasing altitude on exposed ground, growing severity of climate reduces the stature of heather until above 2,500–2,800 feet it forms a dense, flattened carpet growing away from the prevailing wind. Other dwarf mountain shrubs appear plentifully, e.g. bearberry, *Arctostaphylos uva-ursi*, *Empetrum hermaphroditum* and dwarf azalea, *Loiseleuria procumbens*. Locally, 'reindeer moss' lichens, *Cladonia sylvatica* and *C. rangiferina*, and others such as *Cetraria nivalis* and *Alectoria sarmentosa* almost exclude the dwarf shrubs, forming pale yellowish to grey carpets. These lichen heaths are the Scottish equivalents of the vast areas of lichen-dominated vegetation so characteristic of the continental mountain regions of Scandinavia and the Arctic generally, where they provide reindeer with their main food. On less exposed ground with longer snowcover, at over 2,500 feet, the dwarf *Calluna* heath is replaced by *Vaccinium–Empetrum* heath, also lichen-rich in places.

Dwarf *Calluna* heath changes above 3,000–3,300 feet on some tops and spurs to a carpet of woolly fringe moss, *Rhacomitrium lanuginosum*, with mountain sedge, *Carex bigelowii*, dwarfed *Vaccinium myrtillus* and cowberry, *V. vitis-idaea*, grasses such as viviparous sheep's fescue, *Festuca vivipara*, wavy hair grass, *Deschampsia flexuosa*, and bent, *Agrostis canina*, and lichens, notably *Cladonia sylvatica*, *C. rangiferina* and *C. uncialis*. On rocky ground *Empetrum hermaphroditum* is often dominant and locally there is an abundance of least willow, *Salix herbacea*, alpine ladies' mantle, *Alchemilla alpina*, *Festuca vivipara* and *Deschampsia flexuosa*. While there are extensive areas on the western outliers, such as Carn Ban Mor (3,443 feet) and Monadh Mor (3,651 feet) at 3,000–3,600 feet *Rhacomitrium* heath is less continuous on the main granite summits forming the central and eastern Cairngorms.

Also on wind exposed ground above 3,000–3,300 feet is an open tussocky growth of three-pointed rush, *Juncus trifidus*. Moderately exposed ground has discontinuous *Rhacomitrium* heath with *J. trifidus*. Where powder snow lies deeper in winter there are patches of lichen crust with *Cladonia gracilis*, *C. pyxidata*, *C. sylvatica*, *C. uncialis*, *C. bellidiflora* and *Cetraria islandica*, and a little *Rhacomitrium*, with tufts of *J. trifidus* and *Carex bigelowii*. On the most exposed and unstable ground are numerous tussocks of *J. trifidus* with a few lichens. Rare plants include curved woodrush, *Luzula arcuata*, and wavy meadow grass, *Poa flexuosa*, while spiked woodrush, *Luzula spicata*, and dwarf cudweed, *Gnaphalium supinum*, are rather more plentiful. Thinly scattered in the *J. trifidus* heaths, or in bare gravel, are patches of the cushion pink, *Silene acaulis*. This zone of *J. trifidus* heaths extends well over 4,000 feet and is characterised by extremely open ground with the vegetation mostly forming a mosaic or network on areas of predominantly bare gravel, stones and boulders. Some of the great block litters are almost devoid of vegetation except mosses and lichens.

With increasing length of snow lie mat grass, *Nardus stricta* replaces the *Vaccinium* and *Empetrum* heaths and is often accompanied by deer sedge, *Scirpus caespitosus*. Still longer-lasting snow beds show a greater prominence of mosses and liverworts, and often the ground is rather open, showing much bare soil. Species such as *Salix herbacea*, *Alchemilla alpina*, sibbaldia, *Sibbaldia procumbens*, *Gnaphalium supinum*, starry saxifrage, *Saxifraga stellaris*, tufted hair grass, *Deschampsia caespitosa*, *Luzula spicata* and *Carex bigelowii* occur here. Block litters where snow lies long have growths of alpine lady fern, *Athyrium alpestre*, and parsley fern, *Cryptogramma crispa*. The latest snow beds of all have carpets of the mosses, *Dicranum starkei* and *Polytrichum norvegicum*, with few vascular plants. Prolonged melt from these snow drifts ensures a copious supply of water, and there are associated cold springs and rills with spongy green, grey, brown and red cushions of moss and liverwort.

These late snow bed communities occur especially in distinct hollows and in deep, elevated corries. Where snow lies less deeply on flat or gently sloping ground among the *Rhacomitrium* and *J. trifidus* heaths, there is great abundance of *Carex bigelowii*, growing in carpets of the mosses *Dicranum fuscescens* or *Polytrichum alpinum*, but *Nardus* beds take over where snow cover is still more prolonged. Another community here consists of a rubbery skin of small liverworts, especially *Gymnomitrium varians* and *G. concinnatum*, with abundant *Salix herbacea*.

The Cairngorm tableland carries the most extensive and longest-lasting snowfields in Britain, and a few semi-permanent snow beds seldom melt out completely. Climatic severity on the high ground is also reflected in the general instability of the surface deposits. Frost-shattering over a long period has produced a deep layer of rock debris, chemical weathering is slow, and wind removes the finer particles, so that high level soils are raw, shallow and coarse, with much gravel and sand. Frost-heaving and wind action disrupt the vegetation, but there is also recolonisation of bare ground to restore a closed community. The dwarf *Calluna* heaths of windswept spurs are often banded, with strips of vegetation and bare soil alternating along the contours; and in places, large scale solifluction terraces give an interesting vegetation pattern as the shelter/exposure balance changes.

On badly drained flat to gently sloping ground, heather moor passes into blanket bog, with a mixture of *Calluna* and *Scirpus* on shallower peat, and cotton grass, *Eriophorum vaginatum*, and cloudberry, *Rubus chamaemorus*, where the peat is deeper; bog mosses, *Sphagnum* spp., may be abundant in both types. Blanket bog occurs on watersheds up to 3,500 feet and grades into other montane communities of drier ground. Some high level bogs are much eroded and amongst the peat haggs, large pine stumps are frequent up to 2,000 feet; some have been found 500 feet higher, indicating a tree line once higher than it is to-day. Channels, hollows and basins where water seeps laterally have flush bogs with *Sphagna* and sedges; and at high levels they are fed by springs and flushes with spongy masses of mosses and liverworts containing small alpine plants such as *Saxifraga stellaris*, alpine willow-herb, *Epilobium anagallidifolium*, and, more rarely starwort mouse-ear chickweed, *Cerastium cerastoides*, and alpine foxtail, *Alopecurus alpinus*.

Dotterel nest mostly in the *Rhacomitrium–Carex bigelowii* and *Juncus trifidus* heaths on level or gently sloping ground. The nest is often on extremely bare ground, but a vegetated patch big enough to contain the scrape is selected. Some nests are in grassier types of *Rhacomitrium* heath or *Carex bigelowii-Dicranum fuscescens* swards, and a few are in prostrate *Calluna* and lichen-rich *Vaccinium-Empetrum* heaths.

The Central Grampians and Monadhliath

The central Grampians include the hills flanking the Pass of Drumochter and the tops of the Forest of Atholl to the east, forming a great dissected plateau, at 2,800–3,000 feet. These are rounded hills, steep sided locally, but with few corries and cliffs, and more akin to elevated moorlands. The highest summits are Beinn Udlamain (3,306 feet) in Drumochter and Beinn Dearg (3,304 feet) in Atholl. The Monadhliath are another vast plateau-land, north of the Spey, with a general elevation of 2,600–3,000 feet and still more resemblance to moorland, with steep-sided glens only round the outer edges of the massif. The highest tops are Carn Ban (3,087 feet) and A'Chailleach (3,045 feet) above Glen Banchor. The prevailing rocks are schists, gneisses and quartzites of the Moine Series, changing to the Dalradian series from Glen Tilt eastwards, and there are local granite intrusions.

Compared with the Cairngorms, though, these lower hills have a much lesser extent of late snow bed vegetation, and the zone of *Juncus trifidus* heaths is much reduced or absent. Many of the high plateaux have extensive areas of blanket bog, showing variable amounts of erosion, but the main contrast, compared with the Cairngorms, is in the vastly greater extent of *Rhacomitrium* heath. On many tops, ridges and upper spurs at over 2,600 feet, the ground is carpeted over hundreds of acres with a continuous deep and yielding layer of fringe moss, and with only the occasional stone showing above the surface. Sometimes the moss carpet has a billiard table smoothness, but often the ground is thrown into innumerable hummocks, each about 3–5 feet across, and 6–20 ins. high. On sloping ground the hummocks elongate and merge to form low ridges running downhill. These hummocks and ridges are frost-thaw features regarded as the 'fossilised' remains of stone polygons and stripes from an earlier period. In Arctic regions these forms are often a prevailing feature of ice-free terrain.

On some tops there is more stony ground, either with a continuous carpet of moss, through which stones project, or with a more fragmentary vegetation cover and much bare soil and gravel. Such ground has sparse *Juncus trifidus*, *Salix herbacea*, *Gnaphalium supinum*, *Luzula spicata*, *Festuca vivipara* and *Deschampsia flexuosa*, which are all less plentiful in the closed moss carpets.

Dotterel nest on both the closed *Rhacomitrium* carpets and the more open and varied communities of less stable ground. When the ground is hummocky, the nest is placed on top of a hummock, for deer and sheep walk in the hollows.

The Eastern Grampians

The Lochnagar-Clova massif east of the Glen Shee–Braemar road is the easternmost area of high mountain in Scotland, and shows some of the most continental features. From the highest point, on Lochnagar (3,786 feet) a broad watershed lying mostly above 3,000 feet runs south-westwards to the rounded summit of Glas Maol (3,502 feet) above the corrie of Caenlochan, celebrated for its wealth of Arctic-alpine plants. High watersheds at 2,700–3,100 feet above Glen Clova have other flat or rounded tops.

Lochnagar is a granite mountain with vegetation similar to that of the Cairngorms, including *Juncus trifidus–Rhacomitrium* heath on the highest ground. The more southerly hills, around Caenlochan and Clova, are schistose (Dalradian) and here *J. trifidus* is more sparse. The notable features on the dotterel ground are the much reduced extent of *Rhacomitrium* heath, and its partial replacement by *Carex bigelowii*, *Dicranum fuscescens* heath or by lichen-rich *Vaccinium-Empetrum* heath with dense carpets of the 'reindeer moss' lichens. Dotterel here nest in all four of these communities, and sometimes also on variants with more grasses and other small herbs. Soil hummocks and ridges similar to those of Drumochter are well developed in places, and here also nests are sometimes placed on these.

The Western Grampians

To the west of Loch Ericht lie the three high plateaux of Ben Alder

5. Two studies of a cock dotterel's distraction displays. *Above:* The bird shuffles away from the nest with wings beating and white-tipped tail trailing. *Below:* Hump-backed and with tail depressed and fan-like this cock resembles a small mammal.

6. Most hen dotterels take little part in incubation. *Above:* Hen laying her second egg. Through her open beak come soft rippling cries. *Below:* Probably the first photograph taken in Britain of a hen brooding a complete clutch.

(3,754 feet) Aonach Mor (3,650 feet) immediately north, and Creag Meagaidh (3,700 feet) above Laggan. These three plateaux are dotterel haunts, with a greater variety of high level vegetation than the rather lower central Grampians, though they are formed of the same Moine rocks. The still higher tops of the Ben Nevis range farther west, and the peaks of Mamore Forest and Glen Coe are mostly unsuitable terrain for dotterel.

Rhacomitrium heath is widespread on these tops and extensive on the eastern end of the Creag Meagaidh range at 2,700–3,300 feet, but at over 3,300 feet is widely replaced by other communities. *Juncus trifidus* is locally plentiful in the fringe moss carpets, and there is a more open, tussocky *J. trifidus* heath on stony quartzite areas. *Carex bigelowii* heath with *Dicranum fuscescens* or *Polytrichum alpinum* is also well developed. The bare-looking *Gymnomitrium* crust occurs, and there are extensive systems of springs and wide stony flushes on the summit areas. An unusual community is a dense growth of the coarse tufted hair grass, *Deschampsia caespitosa*, and this occurs on all three tops.

The reported dotterel nests in the western Grampians would appear to be in *Rhacomitrium* heath or *Carex bigelowii* sward, but one photograph showed an open growth of *Deschampsia caespitosa* beside a nest.

The Southern Grampians

This district contains the mountains of Breadalbane, with the long range from Ben Lawers (3,984 feet) to Ben Heasgarnich (3,530 feet) on the south side of Glen Lyon and the parallel system from Carn Mairg (3,419 feet) to Beinn a'Chreachain (3,540 feet) to the north. South of L. Tay, the lower hills around Glen Almond and the Sma' Glen have Ben Chonzie (3,053 feet) as their highest point, whilst the higher peaks towards Crianlarich are mostly unsuitable ground.

This is the richest area in the British Isles for alpine plants, which grow in great variety and profusion on the calcareous Dalradian schists of the Breadalbane mountains; many species have their headquarters here, and some are found nowhere else in Britain. This famous flora is, however found mainly on the cliffs and steep slopes below the watersheds, or in springs, rills, flushes and marshes. The rare lesser alpine pearl wort, *Sagina intermedia*, grows on open ground near the top of Ben Lawers and there is much *Silene acaulis* on Ben Heasgarnich summit, but the vegetation of exposed watersheds is generally similar to that of the central Grampians. *Rhacomitrium* heath is extensive and often contains much *Alchemilla alpina*. Some high slopes have irregular systems of solifluction terraces, with *Salix herbacea-Gymnomitrium* crust well represented. Patches of *Carex bigelowii-Dicranum fuscescens* heath occur, and lower down there is a good deal of *Vaccinium-Empetrum* heath often with abundance of grasses, mosses and *Alchemilla alpina.*

There is little information about the nesting places of dotterel in this district, but these are almost certainly within the range of vegetation types described.

The Northern Highlands

Few dotterel appear to nest north of the Great Glen nowadays. Some of the high hills in the west of this region have summit ridges too sharp and narrow for the bird, but there are many suitable tops and high ridges from west Inverness-shire to Sutherland, in the Affric-Cannich hills (Carn Eige, 3,877 feet) and Sgurr na Lapaich (3,773 feet), Glen Strathfarrar (Sgurr a' Choire Ghlais, 3,554 feet) and Monar Forest (Sgurr a' Chaorachain, 3,452 feet), Ben Wyvis (3,443 feet) the Fionn Loch area (A' Mhaighdean, 3,200 feet) and the Fannichs (Sgurr Mor Fannich, 3,637 feet), the Beinn Dearg (3,547 feet) and Seana Bhraigh (3,040 feet) massif, the Ben More Assynt (3,273 feet) group, the Reay Forest (Foinaven, 2,980 feet), Ben Klibreck (3,154 feet), Ben Hope (3,040 feet) and on many lesser summits. These mountains are variously composed of Moine schists and as these are all mainly hard, non-calcareous rocks, underlying geology seems to make relatively little difference to the vegetation of the high ground, which is essentially similar throughout the northern Highlands, except on Ben Wyvis, the most massive of these hills.

Differences in mountain vegetation, both within the northern High-lands and compared with the Grampians and Cairngorms, are evidently related to increasing oceanicity of climate in a northerly and westerly direction. The most pronounced trend is a downward shift in altitudinal zonation of vegetation. Alpine dwarf *Calluna* heath fades out at 3,000–3,300 feet in the Cairngorms, 2,500 feet on the Ross mountains, 2,000 feet in Sutherland and at 1,000–1,500 feet in the Parphe, close to the windswept north-west Sutherland coast. Another oceanic feature is the increasing abundance of *Rhacomitrium lanuginosum*, which in the north-west becomes locally dominant in a wider variety of situations, from drying bogs and heavily burned *Scirpus* moraine ground at low levels to montane dwarf *Calluna-Arctous alpina* heaths, as well as on the mountain tops and in rocky situations generally.

Two trends become increasingly apparent in the high level com-munities in a north-westerly direction. Ben Wyvis has only the ordinary *Rhacomitrium* heath (covering a very large area) but from the Affric-Cannich hills northwards this community contains an increasing abundance of dense patches of thrift, *Armeria maritima*, mossy cyphel,

Cherleria sedoides, and *Silene acaulis*. On many higher hills in west Ross and Sutherland, all three of these alpine herbs occur in profusion irrespective of the underlying type of rock. Where there is intermittent water seepage or on hills with more generally fertile soils, the patches of these plants are much larger, and there is a greater variety of other alpine species such as alpine meadow rue, *Thalictrum alpinum*, viviparous bistort, *Polygonum viviparum*, dwarfed alpine saw-wort, *Saussurea alpina*, *Sibbaldia procumbens* and *Alchemilla alpina*.

In parallel, there is a great increase in the extent of wind-eroded ground and active solifluction on the higher tops towards the north-west. Erosion surfaces cover many acres on some high watersheds and are characterised by a litter of stones lying on bare mineral soil, with an extremely sparse growth of plants such as *Juncus trifidus*, sheep's fescue, *Festuca ovina*, Mountain everlasting, *Antennaria dioica*, *Gnaphalium supinum*, *Salix herbacea* and the lichen, *Solorina crocea*. Solifluction terraces are especially well developed on the Moine hills, and many high slopes show regular 'giants' staircase' systems extending over a vertical distance of several hundred feet.

Really late snow beds and associated moss heaths appear not to extend farther north than Beinn Dearg at present. However, *Nardus communities* and *Rhytidiadelphus* moss heaths are widespread and locally extensive as far north as the Reay Forest, and there is a good deal of a mixed community with *Nardus*, *Rhacomitrium*, *Vaccinium myrtillus* and *Empetrum hermaphroditum*, which appears to reflect longer than average snow cover in the northern Highlands. Some watershed *Rhacomitrium* heaths have much rather sparse *Nardus*.

Although nesting of the dotterel is unknown in the Hebrides, Orkney and Shetland, there is plenty of suitable ground, albeit at a lower elevation than on the mainland. Rhum, Skye and Harris have much *Rhacomitrium* heath (especially the *Nardus*-rich type) on flat spurs and summits at 1,500–2,500 feet, and dwarf *Calluna* heath is widespread at rather lower elevations. On Hoy in Orkney, prostrate dwarf shrub heath with *Arctous* and *Loiseleuria* occurs down to 400 feet, and lichen dominated *Vaccinium-Empetrum* heath at 1,100 feet, on St John's Head; while on Ronas Hill, Shetland, *Rhacomitrium* heath is extensive at 900–1,000 feet, and the summit at 1,450 feet has even more barren granite boulder fields and gravel spreads than the Cairngorm summits.

Of three dotterel nests in Ross, two were in dense *Rhacomitrium-Carex bigelowii* heath and the other was on a more open, partially terraced slope in which the *Rhacomitrium* was grown with *Silene acaulis* and *Alchemilla alpina*. The recent Sutherland nest was on ground with dwarf *Calluna* heath and *Rhacomitrium-Nardus* heath.

Mountains south of the Scottish Highlands

The dotterel breeding areas south of the Highlands are the Southern Uplands of Scotland; the Cheviots, Lakeland fells and Pennines of northern England; and the Cambrian mountains of north Wales. These mountains are lower than the main Highland massifs, and lie in warmer districts; their less alpine mountain top environment is evidently marginal for dotterel at present. These upland districts are similar in their summit vegetation, and a generalised description of the dotterel grounds will be given first.

Most of the 'southern' districts still have heathery grouse moors in places, but mainly within the sub-montane zone below the dotterel's lower breeding limits. Hills rising to 2,500 feet or over are managed as sheep walk and long-continued heavy grazing, combined with repeated moor-burning, has here converted most of the sub-montane and montane dwarf shrub heaths into grassland. The once extensive heaths of *Calluna*, bell heather, *Erica cinerea*, *Vaccinium myrtillus*, *V. vitis-idaea* and common crowberry, *Empetrum nigrum*, on the acidic soils have largely been replaced by grasslands of *Festuca ovina*, *Agrostis canina*, *A. tenuis* and *Deschampsia flexuosa* on drier slopes, or by *Nardus stricta* and heath rush, *Juncus squarrosus*, on damper ground. The southern mountains are thus mainly green and grassy in general appearance, and on the drier deforested lower slopes there are often extensive areas of bracken, *Pteridium aquilinum*.

Few dwarf montane heather carpets remain, for heather is less vigorous and competitive towards its altitudinal limits, and the high level *Calluna* heaths are the first to disappear under management for sheep. Even the remaining areas of dwarf *Calluna* heath are at only 2,000–2,400 feet, well below the probable former limits.

On less severely grazed hills, dwarf shrubs, mainly *Vaccinium myrtillus*, *V. vitis-idaea* and *Empetrum nigrum*, remain dominant in place of *Calluna* above 2,000 feet. Often there is an abundance of 'reindeer moss' lichens, giving locally a vegetation resembling the *Vaccinium-Empetrum* heaths of the eastern Highlands. The crowberry of the southern hills is, however, mainly *Empetrum nigrum*. These *Vaccinium-Empetrum* heaths occupy upper slopes, and where the ground levels out, on ridges and plateaux, a more modified stage has an abundance of dwarfed shoots of *Vaccinium* spp. with patches of *Empetrum nigrum*, in an otherwise grass dominated vegetation. On the most heavily grazed hills, dwarf shrubs are all but eliminated completely, and the slopes and summits at 2,000–2,500 feet mostly have a montane form of the lower level *Festuca-Agrostis* or *Deschampsia flexuosa* grasslands. In this range of alpine heath, grass-heath and grassland, other species may be abundant, such as *Carex bigelowii*, *Salix herbacea*, *Lycopodium alpinum*, *L. selago* and lichens.

Above 2,500 feet on level or gently sloping upper spurs, ridges and summit plateaux, there is a change to *Rhacomitrium lanuginosum* heath, the most montane vegetation type of these southern hills. Though locally extensive, these southern *Rhacomitrium* heaths all show partial invasion and replacement of the moss heath by grasses. On some high tops the *Rhacomitrium* heaths have also been converted almost entirely to montane *Festuca-Agrostis-Deschampsia flexuosa* grassland. Some of the most heavily exploited hills are thus covered from base to summit with acidic grassland. Some tops show disruption of *Rhacomitrium* heath, mainly by frost-heaving and wind, but assisted by the treading of sheep. There are often mosaics of bare soil, gravel and stones, areas with loose or bedded rocks in quantity, and open block litters. Terracing on slopes and surface sorting of debris into networks and stripes is especially marked on ground where the rock is soft and shaly, rather than hard and coarse-grained. Soil hummocks and ridges occur extensively on some high tops, but are mostly smaller and lower than those of many Highland summits.

Where the soil is deeper, there are communities of dense but short *Nardus stricta* and *Juncus squarrosus*, and with increasing stagnation of drainage these in turn change to peat forming vegetation with *Eriophorum vaginatum*. *E. angustifolium*, *Sphagna* and, locally, *Rubus chamaemorus*. Boggy ground often has pools of varying size, and below the highest watersheds there is an emergence of drainage water to give springs, rills, flushes and little marshes. These wet places have vegetation similar to that of comparable situations in the Highlands, but with a lesser representation of alpine species.

The only truly high-level vascular plants of these southern dotterel grounds are *Carex bigelowii* and *Salix herbacea*, and characteristic Highland plants, such as *Juncus trifidus*, *Luzula spicata*, *Gnaphalium supinum*, *Cherleria sedoides* and *Loiseleuria procumbens*, do not occur, whilst others, such as *Silene acaulis* and *Armeria maritima*, are rare or local plants of mountain cliffs here. A few lichens are represented, but even the non-vascular alpine flora of the *Rhacomitrium* heaths is poor.

Because of the artificial spread at higher levels of plants such as *Vaccinium myrtillus* and *Nardus stricta*, it is extremely difficult to determine the influence of prolonged snow-cover on vegetation of mountains south of the Highlands. In the Southern Uplands and Cheviots there are certainly *Vaccinium* and *Nardus* beds which reflect longer than average snow-cover, but farther south the pattern becomes obscured. This lack of clearly defined snow bed vegetation contributes to the rather sub-alpine character of these southern hills.

The following statement on the nesting habitats of the dotterel in Lakeland is attributed to James Cooper (1861), namely, that, 'The birds do not select the summits of the highest mountains, nor do they lay their eggs where the fringe moss grows, but in a depression upon short dense grass, a little below the summit.' This does not tally with more recent experience, as many twentieth-century dotterel nests were on actual summits and in fringe moss carpets. Most nests south of the Highlands are in *Rhacomitrium* heath (with *Festuca ovina*) or in montane *Festuca-Agrostis-Deschampsia flexuosa* grassland (very probably the 'short dense grass' referred to above). Of four nests found during 1959–69, three were in the first type of vegetation and the other in the second; while all four nests seen by E. Blezard in 1925–26 were in the second type. Most nests are on moderately stony ground, with the scrape usually beside one or more bedded stones; as extremes, a nest photographed by J. F. Peters in 1922 was in block litter, whereas one seen by E. Blezard in 1925 was not near stones.

Northern England

Whilst dotterel formerly bred on most mountain groups in Lakeland, the most favoured were those most obviously suitable in terrain, with large summit plateaux and flat-topped spurs. The Lake District mountains are arranged in rather small and compact massifs, separated by ice-steepened U-shaped valleys. Few tops above 2,500 feet are totally unsuitable, but the dotterel grounds here lack the massively extensive character of the Grampians, and suitable terrain on a single top covers scores rather than hundreds of acres. The Scafell (3,210 feet), Helvellyn (3,118 feet), Fairfield (2,863 feet), High Street (2,718 feet) and High White Stones (2,500 feet) ranges all have a good deal of suitable dotterel ground, but on some high flats the ground is too damp and peaty to attract the bird. These Borrowdale Volcanic fells, in particular, are heavily sheep grazed, and the former *Rhacomitrium* heaths have mostly been severely modified though the Scafell range still has fairly good *Rhacomitrium* heath locally. The ground differs between one top and another, as along the Helvellyn range, where Raise (2,889 feet) is bare and stony, Whiteside (2,832 feet) grassier and Stybarrow Dodd (2,756 feet) covered with vegetated soil hummocks.

Skiddaw (3,054 feet) and Grasmoor (2,791 feet), on the Skiddaw Slates, have probably the best areas of *Rhacomitrium* heath remaining but even here the fringe moss carpets are becoming grassy under heavy grazing. Montane grasslands and grass-heaths are well represented down to 2,100 feet on these fells, as on the Buttermere Fells of Robinson

(2,417 feet), Hindscarth (2,835 feet) and Red Pike (2,479 feet), which lack *Rhacomitrium* heath.

Many other Lakeland tops reaching 2,000 feet have the kind of habitat which has in the past attracted, and could still attract, dotterel to nest. Moreover, if old records are accurate, dotterel once nested, at least occasionally, on the much lower limestone uplands of Shap fells, which form a bridge between the Lakeland heights and the Pennines. Areas of limestone pavement and associated close-cropped stony grassland give suitable looking terrain for dotterel, but the elevation is only 1,000–1,300 feet.

By contrast, the Pennines are massive uplands, typically with great sweeps of gently inclined moorland rising gradually to large flat summit plateaux. The higher tops are mostly composed of Yoredale grits and some have a characteristic 'mesa' form, with steep flanking screes below, but others are ill-defined. The northernmost 'Alston Block' between Stainmore and the Tyne Gap contains the highest ground in the Pennines, with Crossfell (2,930 feet), Great and Little Dun Fells (2,780 feet and 2,761 feet), Knock Fell (2,604 feet) and Mickle Fell (2,591 feet) and Little Fell (2,446 feet).

Rhacomitrium-Festuca heath is extensive only on the large, almost level summit of Crossfell, which has well developed soil hummocks, and stone nets and stripes of periglacial type. Montane *Festuca* grassland and *Festuca-Vaccinium* heath are widespread, and from about 2,100 feet upwards are often lichen-rich. Carboniferous limestone outcrops locally and gives a richer *Festuca-Agrostis* turf, but mainly on steeper ground. While there is a considerable area of suitable dotterel ground in this massif, high level peat bog and wet grasslands are interspersed with montane communities of dry ground, and elsewhere in the Alston Block the higher watersheds are mostly covered with deep blanket bog.

South of Stainmore, on the Westmorland-Yorkshire border, and in the Craven Pennines beyond, most of the highest summits lie at 2,200–2,300 feet, and there are large areas of plateau-land at the general elevation of 2,000–2,100 feet. These altitudes are too low for *Rhacomitrium* heath, but in places there are good areas of alpine grassland or *Vaccinium* heath, usually stone-littered in varying degree and often lichen-rich. On some broad watersheds extensive erosion of blanket bog has left a barren terrain in which the exposed sand, gravel, stones and residual peat show variable recolonisation to give stony grasslands and moss heaths of alpine appearance. While the history of the dotterel as a breeding species in this district is completely vague, there is much suitable-looking ground, albeit at a rather low elevation.

The Cheviots

The Cheviots are essentially a moorland tract, rising above 2,000 feet only at their north-eastern end. The culminating point, the Cheviot itself (2,676 feet) has a large, flat top covered almost entirely with deep blanket bog, and only on spurs away from the highest ground, are there small patches of montane grass heath which look at all attractive to dotterel. The few other tops exceeding 2,000 feet are mainly peaty and so cannot be regarded as suitable ground. It is thus not surprising that the history of dotterel in the Cheviots is so scanty.

The Southern Uplands

This extensive tract of mountain and moorland contains two districts with ground above 2,500 feet. In Galloway, the highest hills are formed mainly of Silurian grits, greywackes and shales; the Merrick range contains the Merrick (2,765 feet), Kirriereoch Hill (2,565 feet) and Shalloch on Minnoch (2,520 feet) and the Kells Range has Corserine (2,669 feet), Carlin's Cairn (2,650 feet) and Meikle Millyea (2,446 feet). These mostly have broad, rounded tops and ridges, falling away into steep slopes or deep corries below.

Above 2,400 feet there are good areas of stone-littered *Rhacomitrium-Festuca* heath and montane *Festuca* and *Festuca-Vaccinium* grassheath are still more widespread. Damper ground on the tops has areas of *Nardus* or *Juncus squarrosus* and these species cover large expanses at lower levels. The outlying granite hills of Cairnsmore of Carsphairn (2,614 feet) to the north-east and Cairnsmore of Fleet (2,329 feet) in the south of Galloway both have areas of *Rhacomitrium* heath as well.

The counties of Dumfriesshire, Peebles and Selkirk meet on a high watershed with much ground above 2,500 feet, including White Coomb (2,695 feet), Hart Fell (2,651 feet) and Lochcraig Head (2,625 feet). A few miles farther north, another high watershed contains Broad Law (2,756 feet), Cramalt Craig (2,723 feet) and Dollar Law (2,682 feet). These Moffat and Tweedsmuir Hills lie entirely on Silurian greywackes and shales and have steep, smooth slopes, with deeply cut valleys, flattening out above into broad, rounded or level summits. Only in the glens draining to Moffat Water are there deep corries and cliffs. The summit topography approaches that of the Grampian dotterel haunts more closely than does any other 'southern' hill area. *Rhacomitrium-Festuca* heath is extensive and some tops have mainly the montane *Festuca* grassland. There are extensive areas of soil hummocks locally and the ground is seldom stony or eroded. Some tops have much

Nardus and *Juncus squarrosus*, and some high level *Nardus* patches are associated with prolonged snow cover.

North Wales

The mountains of Caernarvonshire and Merioneth are similar in topography and geology to those of Lakeland. This is a district of steep, precipitous and rather compact mountain systems carved by ice action out of mainly hard igneous rocks, but with softer sedimentary slates locally. The summits and upper ridges of Snowdon (3,560 feet) are sharp and narrow, but in the northern massif dominated by Carnedd Llewelyn (3,485 feet) and Carnedd Dafydd (3,427 feet) there are numerous rounded tops and broad watersheds at over 2,500 feet, and the neighbouring Glyder range also has flat tops such as Glyder Fawr (3,427 feet) and Mynydd Perfedd (2,695 feet). In southern Merioneth, the ranges of Cader Idris (2,927 feet) and Aran Mawddwy (2,970 feet) have areas of level or gently sloping watershed above 2,500 feet. The broad summit ridges of the Berwyn Mountains farther east rise to 2,713 feet on Moel Sych.

On the whole these North Wales tops have stony ground and some show a good deal of bare soil and gravel produced by frost-heaving and erosion. There is a good deal of *Rhacomitrium-Festuca* heath, montane *Festuca* grassland and *Festuca-Vaccinium* grass heaths, but *Nardus* and *Juncus squarrosus* communities are also locally extensive on the high watersheds. The climate, vegetation and terrain of these mountains are so similar to those of Lakeland that the scanty history of the dotterel in North Wales is a mystery requiring explanation other than in terms of lack of suitable habitat. This is a district which could possibly be colonised by nesting dotterel during a period of greater cold.

Historical reflections

The ptarmigan is a truly relict species, in that it is a sedentary creature whose present distribution is the result of a long period of slow change, in which migration was followed by restriction and fragmentation of range, in response to increasingly unfavourable climatic shift. This bird has disappeared from mountain districts south of the Highlands, and is prevented by its limited powers of spread from returning to these areas, where conditions would seem to be suitable at present. The ptarmigan is, nevertheless, widespread in the Highlands, and occurs on virtually all the higher mountain ranges, as well as on lower hills in the far north-west. Possibly, too, heavy sheep grazing in southern mountain districts has rendered these hills unsuitable for the bird by greatly reducing the extent of alpine dwarf shrub heaths. The snow bunting is migratory, so that although it is an extremely sparse nester in Britain at present, it cannot be regarded as 'relict' in the usual sense. This bird is, however, one of the best examples of a 'fringe' species in this country, which now lies on the extreme southern edge of its zone of wide distribution in the Arctic, and has conditions at present evidently marginal for the species.

The dotterel differs from both ptarmigan and snow bunting in its geographical relationships. It is a migratory bird, but its breeding distribution both in Britain and in the world has the characteristically disjunct pattern of a truly relict species, in that it is absent from a great many areas where conditions appear to be favourable. Usually, plants develop a relict distribution through fragmentation of a once more continuous range during a period of unfavourable conditions, but inadequate powers of dispersal are also implied. Global relicts may be senescent species, but some species are relict in one region (especially towards the edge of their ranges), though not in another. The dotterel, as a migrant between the Arctic and North Africa, is certainly not limited by its powers of dispersal but yet shows a marked inability to *spread*, i.e. to establish itself as a breeder in areas at present unoccupied. Its potentiality for spread is shown by its appearance as a breeder in Alaska, but its extreme localisation on the British mountains points to an essential conservatism in distribution maintained over a considerable period.

This conservatism is reflected in the anomalous absence or scarcity of the dotterel in certain mountain areas where the terrain and vegetation appear eminently suitable. Subtle differences, not obvious to the human eye, may make a place suitable or unsuitable, yet it is difficult to understand in environmental terms why the northern England mountains should be more attractive to the bird than those of Snowdonia. The latter are farther south, but some of their most favourable ground lies at a greater elevation, so that there is no effective climatic difference. Nor is it obvious why certain Highland mountains, such as Ben Wyvis and the Sutherland hills, should be so much less favoured than others of basically similar character. The dotterel appears during migration on hills where it does not nest at present, so that fixity of migration routes can hardly be the limiting factor in these localities, which simply remain unattractive in that birds arriving there are not stimulated to stay and breed. This species may have a very strong tendency to return to its birthplace to breed, combined with a somewhat gregarious inclination. This could explain why, in years of high numbers, dotterel occur more numerously in their favourite haunts, rather than become more widely spread over the mountains as a whole.

Whatever the causes, the dotterel is evidently a relatively unsuccessful species whose breeding distribution is determined only partly by the occurrence of suitable nesting places. In this sense it is a relict species with a problematical future status. There is no telling how long this state has existed, but it suggests that the dotterel may not have been as widespread in Britain during the Late-glacial and early Post-glacial periods as the ptarmigan and snow bunting are likely to have been. However, the colonisation of the Dutch polders throws new light on the dotterel's powers of survival and indicates an unsuspected resilience as a species. There has not yet been nesting in the Breckland of East Anglia, where in general appearance the vegetation is locally rather like that of many mountains, with lichens, mosses, sparse grass and bare sand or stones, but the bird passes through this region on spring migration, and such an event would not be beyond the bounds of possibility.

At present, the dotterel's range of adaptation to nesting habitat is much more restricted than that of, say, the golden plover, with which it is often associated. The golden plover is much more numerous and more widespread, both ecologically and geographically, and is thus a more successful species in evolutionary terms. Yet it is far from clear whether or not the dotterel's lack of success is attributable to the narrowness of its adaptation to habitat, or whether this is simply an effect of a reproductive biology which in some way is inadequate for expansion.

Botanical comparisons

It is interesting to compare the distribution and status of the dotterel with those of plants which belong to a parallel geographical and ecological group. Since most central European breeding occurrences of the dotterel appear to be sporadic, it is best to regard this as an Arctic species. Most of the montane plants of Britain are Arctic-alpine, and only a small number occur in the Arctic but not in the Alps.

The mountain plant with which the dotterel is most constantly associated in Britain is *Carex bigelowii*, for a high proportion of nests is placed in vegetation containing this species, and its leaves are often used as nest lining material. Other montane species with a frequent association are *Juncus trifidus*, *Salix herbacea* and *Silene acaulis*. All four species are Arctic-alpine, and have a much wider distribution in Britain than the bird, so that the total geographical similarity with dotterel distribution is limited. Plants which resemble the dotterel in having a patchy and relict distribution in the British mountains, with isolated occurrences in northern England and the northern Highlands, and a main stronghold in the central and eastern Highlands, are the

grasses *Alopecurus alpinus* and *Phleum alpinum* which, however, grow
mostly in damp ground, and have an extra-British distribution diverging
from that of the bird. Several plants unknown south of the Highlands
in Britain have their headquarters in the central and eastern Highlands,
and appear to be associated with a continental mountain climate, and
usually with prolonged snow-cover, e.g. *Carex lachenalii, C. saxatilis,
C. rariflora, Saxifraga rivularis, Cerastium cerastoides* and *Luzula
arcuata*. Most of these plants are best represented on the Cairngorms
but none of them grows actually in the precise habitat in which dotterel
place their nests.

The British plant which appears to show the best worldwide geographical parallel with the distribution of the dotterel is the woolly
willow, *Salix lanata*, which is in Britain a relict species with a few
scattered colonies, mainly on ungrazed cliff ledges in the Grampians, but
with one station north of the Great Glen. Outside Britain, it is an
Arctic-subarctic species, occurring in Iceland, Norway, north Sweden
and Lapland, north Russia, northern Siberia and Outer Mongolia (the
Altai and Dahuria). The very rare Norwegian mugwort, *Artemisia
norvegica* has a still more relict distribution than the dotterel. In
Britain, it is known from two mountains in west Ross, and there grows
at 2,100–2,850 feet on high, stony and mossy spurs which could hold
dotterel too. Elsewhere it is known only in very restricted areas of
Norway, mainly the Dovrefjell, and in the Ural Mountains.

The dotterel's predilection for *Rhacomitrium* heath as a nesting
habitat in Britain is evidently a local feature. *Rhacomitrium lanuginosum*
has a very wide though discontinuous world distribution, but vegetation
dominated by this moss appears to be restricted to the cool oceanic
regions such as the British Isles, Faeroes, Iceland and extreme western
Norway. Its place is taken in more continental regions by lichen heath,
which is a common nesting habitat of the bird in Scandinavia. The
dotterel is evidently adapted to a ground with a certain physiognomy
rather than a particular vegetation – an extremely low growth of mosses,
lichens, short grasses, other small alpine herbs and dwarf shrubs,
ranging from a continuous carpet to a sparse scatter on ground with
little but bare stones and soil. The Altipiano of Monte Maiella, breeding
place of dotterel in the Italian Apennines, is a windswept limestone
plateau covered largely with bare rubble and only a very open mosaic of
vegetation, consisting of varying sized patches, or individual rosettes
and cushions, of alpine herbs and dwarf shrubs. The dotterel has thus
exploited one of the harshest environments on earth, where severity of
climate, as expressed especially in extremes of wind and cold, comes
close to inhibiting the growth of plants altogether.

ACKNOWLEDGMENT

My thanks are due to Donald N. McVean for sharing with me his fund of knowledge regarding vegetation of the Highlands in general and the Cairngorms in particular.

CHAPTER 17

Food and Feeding

IN summer we watch dotterels taking insects and their larvae, but seldom know precisely what they are eating. The Old Naturalists were our masters here. They shot nesting mossfools and described what they found in their stomachs. The scholarly Heysham reflected: 'The stomachs I have dissected were all filled with elytera and the remains of small coleopterous insects which in all probability constitute their principal food during the breeding season.'

In 1925 Ernest Blezard sat beside a dotterel's nest and noted down every movement. 'Occasionally it would straighten itself up to snap up and swallow a cranefly from the myriads swarming the fell top.'

Over forty years later another Cumbrian enthusiast laid the sad little corpse of a recently hatched young dotterel on Blezard's desk. After much analysis and detection, Ernest wrote to me:

'A very full stomach showed that the little bird had been busily feeding on insects. Two craneflies – *Nephrotoma* – had been swallowed complete, long wings, long legs and all. While I have seen nesting adults take flies of this kind when they were swarming over the fell top, they make a sizeable mouthful for a downy chick. Beetle fragments, forming the rest of the cram, were mostly finely ground. The larger remains represented four carabids – *Calathus*, one scarabaeid – *Aphodius* and three weevils – *Otiorrhynchus*, all among the small species in their respective families.

'Three coarse, angular pieces of quartz grit, also contained, might be regarded as a surprising digestive aid, the biggest measuring 9 × 5 × 2.5 mm. and the other two about half that size.'

This chick's extremely large intake of insects emphasises the importance of the dotterel's single parent chick-rearing pattern. In food crises or shortages the competition of a second and larger parent might reduce the chick's chances of survival.

The fragments of quartz grit in the chick's stomach are also significant. The hard wing cases of beetles probably require these artificial digestive aids.

In the Cairngorms and Grampians dotterels are also beetle-hunters. In 1953, when those great concentrations were nesting in the central Grampians, I actually discovered several nests by finding the droppings of the brooding birds! In almost all there were bits of wing cases of beetles.

I have also watched dotterels dexterously lift spiders from between stones.

In summer dotterels have several feeding niches. On the ridges and flats where they nest, they pick up *Diptera*, beetles and spiders. When it is running and circling close to its nest a cock dotterel often snatches a cranefly *Tipula* sp? or even a housefly which a warm current of air deposited on the high ground. On the other hand, when the dotterels first return they feed in marshy places and on the edges of snowfields. Later, when the cocks are brooding, groups of 'grass widows' and associates often continue to hunt craneflies and larvae on moist ground beside burns, tarns or snowdrifts, rather than on the higher and more stony nesting ridges above.

Dotterels often feed at night or very early in the morning, flying far to feed close to tarns or in high peat-bogs. In my tent at night I have often heard them tinkling high in the air as they flew across the Lairig Ghru, possibly to seek wet peatlands across the pass.

Years of much winter snow and late thaws like those of 1951 and 1968 inhibit laying and temporarily prevent dotterels exploiting their richest feeding grounds. Adam Watson (1966) emphasises the importance of micro-climate. 'I have sometimes been amazed to see dotterel and ptarmigan chicks at the critical age of 3–5 days running about vigorously the day after a 12 inch snowfall. However, these snowfalls are usually accompanied by drifting, and unless they are of winter severity, which is exceptional, they do not cover the boulders. The large spaces under and among the boulders are usually completely free of snow and the soil there is not frozen. Many insects are active there, even during the winter on mild days, and midges and other *Diptera* crowd into these places after summer snowstorms. Since ptarmigan and dotterel broods are very often found on ground with boulders, and snow buntings regularly haunt these places, food shortage is seldom probably serious after most snowstorms.' Intimate knowledge of nesting habitat and territory is thus important in all crises.

When we first began to watch dotterels I expected to find parents and broods feeding on the wet hollows and on the flats beside burns and rills, but cocks usually lead their chicks to high ridges and mossy carpets where there are so many beetles and soft insects. The dotterels which nested in the Southern Uplands in 1967 also did this. Donald Watson tells me that within three days the cock had led his chicks 350 feet higher up the hill and about 1,000 yards away from the nest.

Meanwhile, hens often still continue to feed in the wet hollows where I had expected to find cocks and broods. Late in summer groups of young and old often feed on grassy edges beside burns and on patches of wet moss as well as on the high ridges.

Besides snatching beetles, spiders and craneflies, dotterels plunge their beaks deep into wet black moss beside streams, while taking cranefly larvae.

There are a few observations from abroad. Professor R. Collett examined some dotterels shot on Dovre Fjell in June, 1871. 'I found coleoptera, chiefly of the genus *Bembidion*, larvae of Elater, *Lumbrici* (earthworms) and fine gravel.' In May 1874 Collett also recorded several leaves of *Salix*, pieces of straw and insects of different kinds and their larvae, and he found gravel. Collett and Bengt Berg thus both discovered that dotterels eat earthworms in summer. This does not happen in the Cairngorms.

By means of marked birds, Rittinghaus found that hens *and* off-duty cocks fed far from the nest. On the night of 11th–12th June, 1959, for example, a marked cock BG was in the company of groups of four to eight other dotterels which were then feeding up to 1½ kms from the nest on which the hen was brooding the eggs. Rittinghaus also discovered that other dotterels fed a long way from the nest in the height of the nesting season and that parents and broods travelled far in search of insects in root-systems of the alpine grasslands. The dotterels on these fells seemingly did not feed on midges.

In Czechoslovakia, F. J. Turček (*in litt*) tells me that no one has studied the breeding ecology of dotterels in the Riesengebirge, but that these insects – mostly beetles – have been recorded on the dotterel's breeding fells: *Otiorrhynchus arcticus, Otiorrhynchus dubius, Hypnoidus riparius, Corymbites cupreus, Chrysomela lichenis, Amara erratica, Carabus silvestris, Corymbites affinis, Pterostichus/Orites negligens, Arpedium brachypterum*. The distribution of some of these insects roughly coincides with the dotterel's distribution in northern Europe.

A. I. Ivanov has kindly described the food of dotterels in the Soviet Union. In summer different kinds of beetles and *Diptera* and their larvae, and occasionally the seeds and berries of crowberry, *Empetrum nigrum*. On migration they sometimes eat worms and molluscs.

G. P. Dementiev and N. A. Gladkov (1951) also mention 'small green leaves' as an item of food and Uspenskii adds that the stomachs of dotterels shot in Novaya Zemlya not only contained the remains of insects, but also large grains of quartz. Pachosskii (1909) records that on passage flights through the Lower Don steppes, dotterels take large quantities of *Othous niger* and *Aniseplia austriaca* beetles and caterpillars of the butterfly *Cledeobia moldavica*. These beetles damage wheat fields and the butterfly harms steppe hay-crops.

On Dutch polders J. F. Sollie tells me that they apparently only eat

7. *Above:* A newly-hatched dotterel struggling out from under its father. Some birds remove their eggshells as they hatch, and then fly away and drop them within 50 yards of the nest. *Below:* John Markham watches a bird at its nest.

8. *Above:* These chicks ran out of the nest when the cock was disturbed. He is now recalling them to the scoop. *Below:* The first photograph of a cock with chick in a potato field in the north-east polder, Holland.

insects and their larvae and that he has never watched them eating any vegetable food. As these dotterels have nested on fields newly sown with potato, beetroot and wheat they probably feed on different groups of insects.

Homing dotterels usually prefer bare or newly ploughed land in England. Many years ago J. E. Harting examined the stomachs of two dotterels shot in Lincolnshire. One, shot on 5th May, contained 'The remains of *Coleoptera*, larvae of *Lepidoptera* (*Polyodon*) and small particles of grit. A second bird, killed on 7 May, contained, 63 wire worms and 2 beetles. Seven dotterels killed in early May 1888 near Garstang, Lancashire, also had stomachs full of wire worms.'

Henry Cookson suggests that the height of the grain crop determines how long dotterels stay in South Lancashire. One trip stayed four days 'in a grub-eaten portion of oats'. In Martonmoss dotterels were seen pattering for food like lapwings.

On the Scottish borders in the nineteenth century, Muirhead recorded that dotterels favoured habitats where 'numerous stones and turf clods offer protection to the small beetles and other insects on which it feeds.'

What did these dotterels eat? E. Blezard (1958) examined a dotterel killed on 17th May, 1891, at Burgh-by-Sands, Cumberland. Its stomach contained fourteen chrysomelid beetles (*Chrysomela staphylea*), weevils *Sitona*, fine grass shreds and twenty-three particles of grit, mostly white quartz. On 17th May, 1891, another gunner killed three dotterels. Their stomachs were crammed with small beetles, *Chrysomela hyperici*.

On their return journeys – from late July onwards – dotterels seldom visit arable land. We thus know less about what they then eat. On 13th September, 1932, a juvenile picked up not far from Carlisle contained two carabid beetles, *Pterostichus*, and nutlets of *Polygonum persicaria* (Blezard).

Several ornithologists described the feeding behaviour of dotterels in their winter quarters in countries south of the Mediterranean. In the nineteenth century Canon Tristram found great flocks of dotterels feeding on a small land snail, *Helix*, 'which clustered on to the bushes and on every straw'.

Richard Meinertzhagen (1954) remarked that dotterels normally fed on insects, but that in Sinai their staple diet was a small and abundant land snail – 'as many as 200 have been taken from a single bird'.

Jourdain (1940) summarised what was then known about the dotterel's food in summer and on migration. 'Mainly insects: chiefly Coleoptera (*Bembidion*, *Silpha*, *Aphodius*, larvae of *Elater* and *Agriotes*, *Dorcadion*, *Timarcha*, *Otiorrhynchus*, *Cleonus*) and Diptera (*Tipulidae*, small flies, etc); also spiders. 'During migration small mollusca occasionally

D.   M recorded (*Planorbis*, and remains of small *Mytilus*), earthworms (*Lumbricus*) and some vegetable matter including seeds of *Empetrum nigrum*.'

I have summarised some accounts of the dotterel's food and feeding behaviour in Table 23. This merely emphasises the amount of work still to be done.

CHAPTER 18

The Dotterel in Britain

In this century, dotterels have nested regularly in Scotland in small numbers in the counties of Perth, Angus, Aberdeen, Banff and Inverness, probably regularly in Ross and Cromarty, occasionally in Kirkcudbright and Sutherland, and lately in Selkirk and Peebles and possibly in Dumfries. This pattern shows little change from that in the previous century.

In England the story is less assuring. The last decade is, however, more hopeful. Pioneering pairs have apparently started to re-colonise several old haunts.

In 1968 a pair of dotterels possibly nested in a hill in Wales. In 1969 a nest with three eggs was found. To safeguard these Welsh dotterels, I can give no further details.

Let me now discuss breeding distribution country by country and county by county.

SCOTLAND

Orkney and Shetland

The dotterel has never bred on Shetland or on Fair Isle. There are, however, a few passage records. In mid-June, 1870, H. L. Saxby (1874) recorded a dotterel in Unst. About 1894 one was recorded at Dunrossness (Evans and Buckley) and in early July, 1898, another dotterel was seen at Calfirth (O. V. Aplin). On 26th September, 1900, a dotterel was also recorded at Scousburgh (Clark and Laidlaw, 1901).

In autumn dotterels are occasionally seen on Fair Isle. Two were mist-netted on 10th September, 1963, and a third seen on the 13th. (R. H. Dennis, 1963.)

In Orkney, Bree (1850) claimed that a pair nested in Hoy in 1850, but he gave no details. E. Balfour (1968) mentions that in 1935 two boys showed him an unusual-looking plover's egg – 'it resembled, but was smaller, than a lapwing's egg'. Eddie Balfour pointed out the hill on which the egg was found. It certainly looked a perfect nesting habitat.

Caithness

William Dunbar (1860) marked up the dotterel as a regular nester.

Harvie-Brown and Buckley (1887), on hearsay, claim that 'half a dozen specimens' were killed in Caithness and that a pair was 'captured' in April, 1867. This is vague and unsatisfactory. There is no acceptable nesting record.

Sutherland

Charles St. John (1849) claimed that dotterels nested on Ben Klibreck, but he never actually found a nest, or saw dotterels in the nesting season. There is, however, a dotterel in the Dunrobin Museum, Golspie, marked 'Ben Klibreck, 18 June 1846'.

In 1960, Ian Pennie wrote: 'I have never seen a dotterel in Sutherland. They may have been on Klibreck in St. John's time, but they certainly do not now nest regularly, if at all. Duncan Murray and Andrew Mackay, formerly stalkers at Loch Choire, occasionally saw dotterels in summer on Ben Klibreck, usually after a stormy spring. I have no further records.' On 9th June, 1969, Ratcliffe searched Ben Klibreck, but saw no dotterels.

In the 1960s, however, dotterels have been seen in summer on Ben Loyal and Ben Hope, but have not yet been proved to breed.

On 15th June, 1967, Margaret Suggate and friends were searching a hill in north Sutherland. 'Suddenly a bird got up a yard or two from my feet and ran away quickly for about 30 yards. Through my binoculars I saw it was a male dotterel. I then began to look for the nest and among small stones I found the two eggs. Meanwhile, the dotterel had returned and started to feign injury.'

Margaret Suggate thus found the first recorded nest in Sutherland. On 2nd August Donald Hulme found this cock dotterel with one well-grown chick. There were two addled eggs in the nest. The hen had thus laid a third egg after the nest was found. On 5th August, while watching two adults, Dr N. Moore saw a merlin apparently flush two other dotterels. 'The second two appeared to be in juvenile plumage.' In 1967, therefore, two pairs possibly successfully nested. In 1968 and 1969 we searched but failed to find dotterels on this hill.

Ross and Cromarty

J. A. Harvie-Brown and H. A. MacPherson (1904) mention that an old forester, Donald Frazer, asserted that dotterels formerly nested on 'Beinn Bheag' near Kinlochewe.

On 13th February, 1872, at the Glasgow Society of Natural History, Robert Gray read 'A communication' from John Bateson of Shieldaig, who also declared that dotterels nested on two different hills and on his estate in Wester Ross. Again no dates or details.

In the nineteenth century dotterels probably nested in Wester Ross, but I have only one acceptable record. On 7th June, 1897, a shepherd took a clutch on Ben Dearg in Braemore Forest. The three eggs are in the British Museum. In the next fifty years no one has *recorded* dotterels nesting in Wester Ross. But in the last twenty years or so, observers have seen dotterels on several Wester Ross bens. An occasional pair probably sometimes nests in the Fannichs. On 22nd July, 1948, Dr H. Milne-Redhead watched a group of four on one summit plateau and on 4th June, 1969, a friend watched a cock land on another hill on which young were seen later in summer. Then, in 1971, Ivan and Mary Hills found a nest with three eggs at 1,500 feet on another hill in the Fannichs.

On 2nd July, 1950, Margaret Swan (1950) watched a very tame dotterel near the top of a big hill (above 3,000 feet) in Wester Ross, and on 15th May, 1954, Adam Watson saw four dotterels – probably two pairs – running on another west Ross high plateau with typical grass-moss vegetation. A local deer stalker had also periodically seen dotterels on this hill.

In mid-July, 1956, Professor W. H. Pearsall saw a party of five dotterels – apparently one adult and four unfledged young – on Beinn Eighe.

On 9th June, 1956, Ratcliffe found two nests on remote hills in west-central Ross. In 1959 he found two broods, and on 10th June, 1968, another nest on the 2,800 foot contour. On 23rd July, 1970, Sandy Tewnion also watched a cock displaying and evidently with chicks at about 3,000 feet.

There are also reports that dotterels still nest on the Beinn Dearg group, but I have no recent breeding records. On 5th June, 1968, however, Ray Collier watched a dotterel flying over a hill 1,500 feet high in Inverlael Forest.

The remote bens of Wester Ross thus provide the dotterel hunter with real challenge.

There are more records of dotterels nesting in Easter Ross. Each year between 1900–04 nests were taken on Ben Wyvis. In 1903 Frederick Courteney Selous, the great lion hunter, took a clutch, and on 11th June, 1904, John Baldwin Young, of Sheffield, also found a nest on Wyvis. Then, on 25th May, 1912, the late Clifford Borrer took a nest on this formidable hill.

I do not know whether dotterels still continue to nest on Ben Wyvis, but Richard Meinertzhagen found nests in 1923 and 1927, and told me that in both years the dotterels reared broods. In 1935 the late Dr John Kennedy found a nest with three eggs on the great mossy plateau, and in early June 1941 Ivan Hills, then in the Seaforth Highlanders, watched a pair on the same ridge, but was 'posted' before finding the nest.

The late Tom Gordon, formerly deer stalker on Wyvis Forest, kindly informed me that until the late 1930s one pair regularly nested on the 'march' between Wyvis and Kildermorie Forests. In 1956, 1959 and 1969, however, Derek Ratcliffe saw no dotterels on this wonderful massif.

I have one record of dotterels seen on a high moine-schist hill between Glens Cannich and Affric. In June 1957 John MacRae, a keen field naturalist, saw a pair on the north-east spur – the only dotterels that he ever saw on the Ross-Inverness hills.

Inverness (North of the Caledonian Canal)

Few dotterels have been recorded in the nesting season in Inverness (north of the Caledonian Canal). In 1954, however, Adam Watson met a hill-climber who had seen dotterels in summer on Sgurr a Choire Ghlas and Sgurr Ruaidh in Glen Strathfarrar. In August 1958 Derek Ratcliffe also saw a dotterel flying over the top of Sgurr-na-Lapaich (3,777 feet) in Glen Cannich.

On 12th June, 1965, M. J. P. Gregory watched an extremely tame dotterel running along a ridge on the 3,250-foot contour of a hill on the Inverness-Ross march. This bird probably had a nest. In 1970 Gregory also reported dotterels in the nesting season on another likely hill in west Inverness.

Isolated pairs also probably sometimes nest on other hills in this rough country which few ornithologists ever explore.

Monadhliath

Few ornithologists have studied dotterels on these Moine-schist hills, west of the Spey. Before the First World War, Dr E. S. Steward found a nest on Carn na Criche and in the late 1940s and early 1950s Neil Usher met with some nesting pairs. On 22nd June, 1953, Richard Perry also saw a solitary cock running on the 2,850 contour of the Geal Charn (3,036 feet), west of Newtonmore; but he could find no nest or brood. Dotterels, however, continue to nest regularly in small numbers.

West and Central Inverness

Dotterels regularly nest in fluctuating numbers on the hills west of
Loch Ericht and south of the Laggan–Spean Bridge road.

Finlay Mackintosh, formerly head stalker at Ardverikie, often
watched dotterels nesting on Ben Alder, but he kept no dated records.
On 5th July, 1950, Dr H. Milne-Redhead watched excited dotterels
on Ben Alder. On 10th May, 1953, Adam Watson, Sen., saw two pairs
on Aonach Beag, south of Spean Bridge, and on 5th August, 1957,
Derek Ratcliffe watched a family group on Aonach Mor. Then, on one
hill in this group, in June, 1968, Ivan and Mary Hills found the nests of
two of the three pairs they located.

A few pairs also annually nest on the knobbly hills of Gaick Forest.

The heartlands of dotterels in Inverness are on the Cairngorms and
in the Drumochter Hills. Three counties, Inverness, Banff and Aberdeen,
share the Cairngorms, and Inverness and Perth divide the hills on both
sides of the Drumochter Pass.

Cairngorms

The west Cairngorms, a range of great hills and high tablelands, lie west
of Lairig Ghru. In the west Cairngorms are Sgoran Dubh Mor (3,625
feet), Sgor Gaoith (3,635 feet), Carn Ban Mor (3,443 feet), Monadh
Mor (3,651 feet), Brae Riach (4,248 feet), Cairn Toul (4,241 feet) and
Beinn Bhrotain (3,795 feet). Sgoran Dubh Mor, Carn Ban Mor, and
Sgor Gaoith are in Inverness, but the great arctic hinterland between
Cairngorm and Ben MacDhui is in Banff. Banff and Aberdeen also
share Ben MacDhui (4,296 feet), the highest hill in the Cairngorms.

In 1786 Colonel Thomas Thornton, on his most ostentatious safari,
gave the first eye-witness account of dotterels in the Cairngorms.

The nineteenth century Victorian naturalists added little to our
knowledge. On 29th May, 1876, however, Major-General Sir David
Bruce took a clutch in Aberdeenshire. This is an interesting record as,
three days later, this gallant knight shot a cock snow bunting on Brae
Riach, thus giving a clue to his hunting ground. Two other snow
bunting hunters, Captain Savile G. Reid and His Pomposity W.
Ogilvie-Grant, of the British Museum's Bird Room, were others who
enjoyed 'the double' in 1893. On 5th June they robbed a snow bunting's
nest on Ben A'an and on the 15th they 'procured' a clutch of dotterel's
eggs on the same hilltop.

From the early 1920s onwards the dotterel's distribution was more
fully studied. During the last fifty years dotterels have nested regularly
in fluctuating numbers on many Cairngorm hills. I have already

described what is known about numbers and fluctuations. Some hills
and ridges certainly attract them more than others. In the 1930s and
1940s, for example, we never found dotterels nesting on one hill where
they had nested in the 1920s. They only seem to favour lower hills in
years of exceptionally heavy snowcover. In the early 1900s, one pair
nested successfully on the Geal Charn (2,692 feet) above the Dorback
Moor.

Central Grampians

The Drumochter Hills, on the marches of Perth and Inverness, have a
long history as dotterel nesting grounds.

In the *Old Statistical Account* the Rev. John Webster declared that
dotterels had become much scarcer 'since the county was improved',
but mentioned Struan and Wemyss as two haunts in Perthshire.

About 1870 Harvie-Brown, who needed dotterels' eggs for his large
collection, started importuning the Drumochter gamekeepers. In 1873,
as we have seen, Feilden and he went and conquered; but they were
merely the first in the queue. Every year the pattern of the dotterel's
breeding distribution in the central Grampians became clearer as empty
spaces in the egg cabinets filled up!

East Grampians

The East Grampians lie in east Perth, south and west Angus, and
south-east Aberdeen. These hills were the stamping ground of William
Scrope, Esquire, and known to all who have read his classic *Days of
Deer Stalking*.

The hills and high ground of Caenlochan Forest, including Monega
(2,997 feet), Caderg (2,950 feet), Glas Maol (3,502 feet), Cairn
of Claise (3,484 feet) and Tolmount (3,142 feet) are in Angus, as are the
hills west and east of Glen Clova, including Driesh (3,115 feet). Carn
an Tuirc, and 5½ miles north-east, the huge massif of Lochnagar
(3,786 feet) in Balmoral Forest, are in south-west Aberdeen.

In the *Old Statistical Account* the worthy ministers of the Parishes of
Carmyle and Inverarity in Angus mentioned that dotterels rested on the
low ground before flying away to the high tops towards the end of April.
They returned early in August, 'after abiding here about three weeks
they fly off southwards and are not seen again until 1st April following.'

Thomas Pennant (1790) did not venture to the high tops, but
mentioned that Lord Fife's factor asserted that dotterels still bred on
the 'higher Braemar hills but less frequently than formerly'. On 8th
August, 1851, the learned Professor William MacGillivray watched a

dotterel which 'pretended lameness' and hovered around near the summit of Lochnagar; but in the next sixty years there were few published records. An old man who lived in Glen Shee between 1830 to 1895 told H. A. MacPherson that, as a young man, he had found dotterels' eggs and downy chicks. Before 1914 the Blackwoods watched dotterels nesting on the Angus hills and in 1912 L. J. Rintoul and E. V. Baxter recorded a nest in the county.

In the last decade, Derek Ratcliffe, G. Sutherland, Sandy Tewnion and Adam Watson have proved that dotterels are nesting regularly on several east Grampian hills. Watson also twice watched cock dotterels in summer on bare and windswept ridges of hills in Glen Esk – one at 2,100 feet and another at 2,300 feet. On 30th June, 1953, Canon G. H. K. Hervey watched a pair above the 3,000 foot contour on a Glen Clova hill. H. MacPherson, deerstalker in the Forest of Atholl, also tells me that during the last fifteen years he has seen many dotterels on the Atholl hills, but had only met with one nesting pair. In 1969 H. Milne-Redhead also recorded dotterels in summer on the Perth–Aberdeen march.

West Perth

Many remote hills in deer forest north of Loch Rannoch probably sometimes harbour nesting dotterels, but I have no first-hand records. Farther south, however, between Lochs Rannoch and Tay, dotterels formerly nested and sometimes still do so.

About 1850 dotterels nested in some numbers above Loch Tay. D. Dewar found nests but seldom shot the parents. In the 1860s E. T. Booth proved that dotterels were nesting in some numbers north of Glen Lyon, but by 1879 they were 'rapidly decreasing'. Harvie-Brown suggested that dotterels were increasing on hills south of Loch Tay. 'A second locality, still further south, remains uncertain.' I do not know precise locations, but dotterels possibly nested north or south of Glen Almond.

In the last sixty years, few have recorded mossfools on the hills of Glen Lyon or west Perth. On Ben Lawers on 29th July, 1908, Blackwood saw a family group with three young flying strongly. This little trip was possibly a Ben Lawers brood. On 14th July, 1960, Dr Jack Dainty also watched a dotterel behaving as if it had young nearby. Ben Lawers is, however, possibly a marginal nesting haunt.

Dotterels nevertheless visit and still occasionally nest in Glen Lyon. On 21st May, 1936, a gamekeeper saw a dotterel on the high ground of Fortingall and in the early 1960s Professor Duncan Poore and another ornithologist also saw dotterels on these hills in summer. In June, 1944, and June, 1946, on the other hand, Dr George Franklin found no dotterels on Meall Garbh, Cairngorm, Schiehallion, Carn Daimh and other apparently suitable hills.

Then, in 1968, dotterels again nested in Glen Lyon – the first acceptable record, to my knowledge, since Booth! On 18th May Mr and Mrs M. J. P. Gregory watched three adult dotterels near the top of a hill in this glen. Then, on 29th July, David Merrie found a cock dotterel and at least one chick, about two to three days old, at 2,800 feet on an adjacent top. In 1968, however, dotterels nested exceptionally late and in low numbers in the central Cairngorms. This was probably due to the backward spring. Dotterels still nest periodically in Glen Lyon. On 18th July, 1971, Patrick Stirling-Aird saw a pair with one fairly wellgrown chick at about 3,300 feet. Investigation of the Sma' Glen could also be revealing. In the context of the current climatic deterioration and the southerly extension of boreal species like redwing and wood sandpiper, these are particularly interesting records.

Southern Scotland

Many naturalists have searched the Southern Uplands but I know of few proven breeding records. In the 1830s Thomas Heysham tried but failed. Perhaps no 'able assistant' there.

Dumfries

Dotterels have probably nested in Dumfries, but I have been unable to discover any authentic records. Sir Robert Jardine asserted that 'one or two pairs nested on the hills close to his estate'. In 1887 Robert Service assured Professor Newton that a few pairs nested between Moffat and Minningaff; but I can find no dated records. Bird watchers in the south of Scotland are a dour and persistent race. In 1924 E. R. Paton paid a watcher to look for dotterels on hills on the Ayr–Dumfries border but he had nothing to show for his money and enterprise.

William Austin (*pers. comm.*) was brutally frank about the Dumfries records. 'In this century I know of no acceptable records of dotterels breeding in the county and the few records in the 19th century are extremely doubtful. Our hill climbers and ornithologists are constantly on the look-out for dotterels, but have never recorded seeing them on any suitable hill in summer.'

Kirkcudbright

Robert Service (1905) asserts that dotterels' eggs have been taken on the hills above Tegior, but he gave no year or dates. Donald Watson has kindly summarised more recent evidence. 'Now and again I get word from shepherds and the like, who have apparently seen dotterels on two of the larger hill ranges in the Stewartry; one shepherd claims to have seen a nest a few years ago on one of these hills. There is also a suggestion that nesting took place on another hill in the 1930s. Arthur Duncan also told me that an egg collector took eggs in Galloway about 1930. I wonder if the early 1930s were a good period for dotterels in this area.'

But there was more to Donald Watson's story. On 30th May, 1967, he himself found a dotterel's nest in the Stewartry and on 25th June watched adult and brood on the move. Since 1967 there have also been several significant but unproved reports of breeding.

Selkirk and Peebles

A stray pair possibly sometimes nests undetected. On 26th May, 1962, for example, an acquaintance came upon a trio of dotterels near the top of a most suitable Southern Upland. My friend searched and searched again, but he found no nest. In 1970, however, a pair nested on another hill quite close to where he had watched the trio. Dotterels are thus now probably coming back to the Southern Uplands where so few have ever found them.

Roxburgh

There is one ancient and dubious record of dotterels breeding in Roxburgh. A Reverend Gentleman wrote in *The Old Statistical Account* asserting that they nested near Oxnam.

ENGLAND

Between 1850 and 1899 dotterels nested in regular but declining numbers in Cumberland and Westmorland, probably irregularly in Yorkshire, and possibly in Lancashire. All acceptable breeding records, however, refer to Cumberland and Westmorland. In Cumberland dotterels nested on Skiddaw, Helvellyn, Robinson, Grasmoor and Crossfell.

In Westmorland they bred on Cockley Fell, Fairfield (Helvellyn), Orton Scar and also probably on fells around Crosby Ravensworth, Kentmere High Street, and on Mardale Fell. 'Bow Fell on the borders of Yorkshire and Westmorland is, so far as I know, the first northerly mountain on which these birds stay to breed.' (J. Watson, 1888.)

Dotterels also possibly nested on the hills at the head of Wyresdale in

Lancashire. Watson reported that a hen 'with her brooding patches bare' was shot one year on 1st August at Beaton Fell and that the Reverend J. D. Banister had known a hen 'to be killed there with eggs in her at maturity'.

In these fifty years I have documented records of twenty-seven nests: twenty-one in Cumberland (Skiddaw eleven, Grasmoor three, Robinson two and Crossfell one) and three in Westmorland (Cockley Fell two and Fairfield, Helvellyn one). The locations of the rest are not known, but were all in Cumberland or Westmorland.

R. Chislett (1952) also mentions that in 1895 young were seen in Yorkshire.

Between 1900 and 1926 dotterels nested sparingly, but probably regularly in England; thenceforward they nested only sporadically.

In the first decade I have reliable records of four nests; three from the Buttermere Fells and one from the Derwent Fells in Cumberland. R. Chislett (1952) also records that in 1905 young dotterels were seen on Yorkshire fells. But in this period dotterels seldom nested in Yorkshire.

Between 1910 and 1926, when so few dotterels were nesting in England, a remarkable group of north-country nest-hunters actually found twenty-eight nests with eggs and thirteen broods of young. They located twenty-one nests and eleven broods in Cumberland; four nests in Westmorland; two nests in Northumberland; and one nest in Durham. I have no acceptable Yorkshire records later than 1905. Table 24 summarises all known records of dotterels in England between 1900 and 1928.

Between 1927 and 1931 I have no reliable breeding records. Thenceforward, between 1932 and 1971, I know of at least fourteen nests, but, until 1969, never more than one nest in any year, and only thrice nests in consecutive years on the same hill. Derek Ratcliffe gives details.

Dotterels were also seen on likely nesting ground in spring or summer on two different fells in 1954, 1964 and 1970, and on one fell in 1931 and 1961. Cumberland, Westmorland, Yorkshire and probably Northumberland share these records.

The 1960s have been particularly encouraging, with at least one nest in 1960 and 1968 and two nests on the same fell in 1969.

WALES

I have referred to the pair which probably nested 'somewhere in Wales' in 1968 and certainly nested there in 1969 – possibly the most exciting British dotterel news of this century. Yet it was not quite unexpected. On 30th April, 1949 – 'a sunny and bright day, warm out of the west wind', a good ornithologist saw a dotterel 'tame as they always are', running on the very top of Foel Fras (3,091 feet) in Caernarvon.

In the 1950s a research student from Bangor University also watched a possible nesting pair in Caernarvon (Ratcliffe *in litt*). On 6th May, 1967, three dotterels were also seen on the Carneddau (P. E. Davis and P. H. Jones, 1967). Dotterels are also sometimes recorded on passage.

Conclusions

What conclusions can we draw from the dotterels' recent distribution in Scotland, England and Wales? In the Cairngorms and central Grampians I doubt whether distribution has greatly altered during this century, but elsewhere I believe that long-term climatic trends have caused changes.

In Britain dotterels are 'fringe birds'. Breeding groups in the west Grampians are the most westerly regular nesting populations in the world. In phases of warmer climate the dotterel's breeding distribution thus tends to contract and in more severe climatic periods to extend. We can expect to find evidence of these climatic influences in outlying or marginal habitats – for instance, north of the Caledonian Canal, west Perth, Southern Uplands and English fells. What do we find? In the last twenty years meteorologists have evidence of wobbling or deteriorating climate. Simultaneously, dotterels are returning to long-deserted haunts. In the 1950s and 1960s they are apparently re-colonising some Lakeland fells. In 1967 dotterels nested in Sutherland and Kirkcudbright and in 1969 a pair bred in Wales.

'Fringe birds' seem sensitive to subtle changes in climate and environment. Between 1966 and 1972, for example, more snow buntings have spent the summer in the Scottish corries than in any period since the nineteenth century. More cocks are singing and possibly more pairs nesting in the high Cairngorms, while others are now prospecting those lower hills where the Old Naturalists record them in severer climatic decades during the late eighteenth and early nineteenth centuries. On 18th May, 1968, for example, M. J. P. Gregory saw a cock on top of a hill in west Perth.

In the late 1950s and 1960s boreal birds – osprey, snowy owl, wood sandpiper, fieldfare, redwing, bluethroat and others – have also pioneered and nested in northern Scotland. The southerly extension in breeding distribution of so many boreal species is unlikely to be fortuitous. We can thus better understand shifts in the dotterel's range against a background of climatic oscillation or change.

CHAPTER 19

The Dotterel as a Breeding Bird in England

By D. A. Ratcliffe

SINCE 1800 a succession of patient and energetic dotterel enthusiasts, at least one to each generation, has left a fund of information, both published and unpublished, about the species in northern England. The earliest account of the dotterel as a nesting bird in England is the celebrated paper by T. C. Heysham in the *Magazine of Natural History* for 1838. Interesting contributions on nesting were later made by Francis Nicholson in MacPherson and Duckworth's *Birds of Cumberland* (1886) and by John Watson in the *Natural History Record* for 1800, but the Rev. H. A. MacPherson, in his *Vertebrate Fauna of Lakeland* (1892), has provided the fullest account of the history of the dotterel in this region up to the end of the nineteenth century.

The first definite nesting record is for a clutch of eggs taken on Skiddaw, Cumberland, in 1784, as recorded by Dr J. Heysham in MS. The younger Heysham remarked in 1830 (*Magazine of Natural History*) that the eggs of the dotterel remained undescribed, although 'they constantly breed in the mountainous parts of Yorkshire, Westmorland and Cumberland,' and he forthwith set himself the task of description, relying on correspondents in dotterel country to secure the necessary specimens. These various agents supplied corpses but no eggs and it was not until 1835 that Heysham achieved his object. The story of the success, as described by Francis Nicholson in MacPherson and Duckworth (1886) is told in Chapter 2. Other breeding records for the Lake District followed during the next few years, and in June, 1841, a nest was found on the adjoining Pennine range in the same counties of Cumberland and Westmorland. The second half of the nineteenth century produced several notable dotterel hunters, who between them found a considerable number of nests, on many different hills; the Keswick naturalist, W. Greenup, had at least ten nests in the 1850s. During the period 1835–1900 nests were recorded at one time or another from nearly all the higher fell-ranges of the Lake District. The massifs dominated by Skiddaw, Grasmoor, Robinson, High Stile, Great Gable, Scafell, High White Stones, Helvellyn, Fairfield and High Street were all favoured haunts; only the Pillar and Coniston Old

Man Groups were not mentioned, yet these, too, have likely ground. A number of Pennine summits, both in the Crossfell range and in Yorkshire, were known as nesting places. Moreover, Watson (1888) stated that dotterel bred at times on the lower uplands which connect the lake fells and the Pennines (Orton Scar and Crosby Ravensworth Fell) as well as on some of the lesser heights of the eastern Lake District, such as Cockley Fell in Longsleddale. It is now impossible to judge the reliability of supposed nesting places from other upland districts much farther south in England, but the dotterel may once have had a much wider breeding distribution in this country, even within the last few hundred years.

It is difficult to gain from the records any idea of the actual size of the dotterel nesting population in northern England during the nineteenth century. Some haunts were evidently favoured more than others, but this may have varied from year to year, and there certainly appear to have been annual fluctuations in the numbers of nesting pairs. Watson (1888) remarked that five or six pairs frequently bred at no great distance from each other, and some of the more extensive tops frequently held at least three nesting pairs in the same season. Yet, from the opinions expressed by the knowledgeable ornithologists of the time and from their recorded evidence, it is clear that the dotterel was never known to be more than a sparse nester in northern England, thinly distributed over a wide area. This is despite the fact that during spring migration, trips of dotterel, each of up to two hundred birds, were then of regular occurrence on the coastal salt marshes, lower moors and high fells of Lakeland. Yet it is said that the trips used to collect on certain mountains before dispersing, so that the larger gatherings may have represented a considerable proportion of the spring arrivals in the whole region, and this may have given the impression of a larger migration than actually took place. There is no telling whether some or all of these birds moved on to other regions, or whether they all stayed in Lakeland, but a hundred dotterel would represent only a thin scattering when distributed evenly over all the likely nesting grounds of northern England. In any event, it would seem likely that during good dotterel years, nesting density in Lakeland was comparable to that of the nesting grounds in the Grampians at present. This would allow us cautiously to estimate the total annual breeding strength as fifty to seventy-five pairs, in good years, up to about 1860.

Even by 1885 experienced observers were stating that the numbers of nesting dotterel had declined considerably, and foresaw the possibility of its final disappearance from the Lake fells. Francis Nicholson recorded that in 1885 and 1887 he found only three pairs despite intensive searching of all the likely ground, but it is doubtful if he really examined

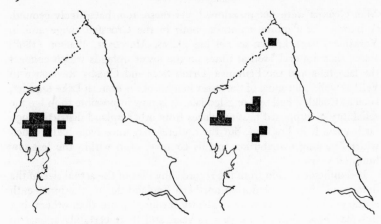

Breeding distribution in England

FIG. 11. In 1850–99:
10 km squares of National Grid
in which Dotterels nested in this
period.

FIG. 12. In 1900–45:
10 km squares of National Grid
in which Dotterels nested in this
period. *Note:* From 1927 onwards Dotterels ceased to nest
regularly in England.

all the possible nesting places during a single season. MacPherson
(1892) records that H. E. Rawson found three pairs nesting in each of
the three years 1889–91, all evidently in Westmorland. Professor J. H.
Salter saw two breeding pairs on the same day in 1894; and up to 1900
the dotterel probably continued to nest regularly on several Lakeland
ranges, though in reduced numbers. In 1896 J. Backhouse said that it
formerly nested on Crossfell, but that none had been seen there during
the previous few years.

Since 1900 the dotterel has certainly been a scarce and elusive nester,
with fewer known breeding haunts than during the previous century,
and nesting records outside Cumberland and Westmorland are
particularly vague. Nevertheless, a few pairs seem to have attempted to
breed during most years in the first quarter of the century, and during
this period J. F. Peters acquired an unsurpassed knowledge of the
dotterel as a Lakeland bird. Peters found nests with eggs or young in
many of the years from 1905 to 1924, and saw birds on nesting grounds
during most years when actual nests eluded him. He even suggested

that dotterel had increased during this time but said that the number which stayed to nest was variable. The records of other observers fill in most of his blank nesting years. One favourite summit had two nests in 1908, two in 1910 and no less than three in 1911, two of these nests being only a hundred yards apart (J. F. Peters), and in the same locality, Peters saw two and possibly three pairs (but no nests) in 1921. A few other fells sometimes held single pairs and on a range which includes several suitable tops there was usually at least one pair, and often two or three, up to 1924. George Bolam found two or three pairs nesting regularly between 1913 and 1926 on one summit of another massif, and on the same extensive breeding grounds Ernest Blezard and Ritson Graham discovered three nests with eggs in 1925. Another hill far distant held a nest in 1925 (W. H. Pearsall).

The year 1927 may be given as the end of the period when dotterel were known to nest annually in the north of England. Since that year only a handful of nests has been reported, and no more than two in any one year. Single nests were found in 1932 by E. Telford and in 1934 by R. H. Brown, while another nest was reported at second hand in 1933. Professor W. H. Pearsall located a nest in 1937 and during the war period 1939–45 Dr E. S. Steward encountered at least one nest, whilst a second was said to be discovered by a gamekeeper. After this, there was a lapse until 1956 when a nest was seen (per R. H. Brown) and subsequently seven nests have been found, four in 1959, 1960, 1968 and 1969 by D. A. Ratcliffe, another in 1969 by R. Laidler, and two in 1970 by D. Clark and R. Stokoe.

While it is almost certain that other discovered nests have not been heard about, and equally likely that some have escaped detection altogether, one may say that the dotterel has probably not been a regular nester in England during the last thirty years or more. During the period 1945–70 I have paid 133 visits, involving several hundred hours' observation time, to known or likely dotterel nesting hills in the north of England. The reward for this effort has been four nests with eggs, a pair of dotterel, three single birds, and a cast feather. As dotterel have been known to breed on at least thirty-two hills in this region, proportional coverage of the available ground has been low (about five hills annually, on average), and these results have only relative meaning. For comparison, fifty-one visits to known dotterel hills in the Scottish Highlands during 1952–70 produced seventeen nests with eggs, three broods of chicks and nine sightings of birds when no nests were found. Moreover, there have been many more searchers besides myself active in northern England during the period since 1945, yet there have been few other reports of nests, or even of birds seen on the fells in nesting time. Even allowing for the probability that some tight sitting birds have been missed, the recent scarcity of dotterel in this region thus appears to be a real one.

There could be several reasons for this decline. To some extent it coincided with the excessive persecution of the species by Victorian anglers for its valued feathers and this has often been blamed as an obvious cause. At this time, parties of gunners regularly scoured the Lake fells and made inroads into the trips. Shooting of dotterel on the nesting grounds had virtually ended by the beginning of the twentieth century, but no recovery followed. Moreover, regular nesting ended nearly three decades after this destruction ceased. A good many nests were robbed by collectors before and after 1900 and this could possibly have had an effect on the breeding population.

However, it is evident that the number of dotterel *arriving* in the region during the spring migration has suffered the same decline as the breeding population. Whereas trips were once a normal occurrence in the low country of north-west England, as well as on the fells, they are hardly ever seen in either situation nowadays. Ernest Blezard told me how the celebrated Solway wildfowler William Nichol (1854–1934) began his fowling career as a boy by stalking trips of dotterel on Skinburness Marsh with a bow and arrows. Yet even by 1892 MacPherson indicated that trips had become infrequent on the coastal salt-marshes and by 1920 they were a decidedly rare event. It is possible that persecution may gradually have reduced the migration stream of dotterel reaching northern England, on the assumption that decline in total population of a species would begin to show first at the edge of its range. Harvie-Brown (1906) claimed an increase in dotterel population and southward extension of range for the Highlands around 1902–5, but records are too few to tell whether this change affected northern England. Peters commented on a possible increase some time after 1900, but any such resurgence in Lakeland could at most have been only temporary, and might have been merely a short run of good years.

There has been growing disturbance on many of the favourite nesting haunts as the tourist traffic in Lakeland has increased and certain of the old nesting grounds are now much frequented by hikers. Some fairly recent sightings of dotterel have named lower tops little frequented by walkers, but at least three of the recent nests (all successful) were in places passed closely by considerable numbers of people during the incubation period. It would therefore seem that the bird can tolerate a moderate amount of casual disturbance and that the greatly increased numbers of people on the fell tops is not to be blamed for the decline. This factor could, however, be limiting beyond a certain level.

Again, there may have been increased predation by natural enemies. As early as 1920, J. F. Peters pointed out that the numbers of carrion

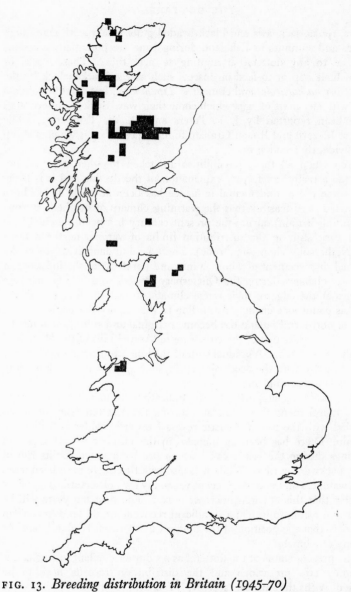

FIG. 13. *Breeding distribution in Britain (1945–70)*
10 km squares of National Grid in which Dotterels were recorded in summer on suitable breeding habitats in Scotland, England and Wales during this period. Nests or broods were not recorded in all these squares.

crows, rooks, jackdaws and black-headed gulls which work many high slopes and summits in Lakeland during June are potentially a serious menace to any dotterel attempting to nest there. These flocks or individuals appear to feed on insects such as the brackenclock beetle, *Phyllopertha horticola*, and craneflies, *Tipula* spp., but would doubtless deal with any nests of eggs which come their way. Sucked dotterel eggs have been reported by J. F. Peters and G. W. Temperley, while Ernest Blezard and Ritson Graham found a nest of eggs which a sheep had evidently trodden in.

Even when all these possibilities are taken together, they do not provide a really satisfactory explanation of the decline and it is likely that some more fundamental influence has been at work. It has been suggested very feasibly that the warming climate of northern Europe (including Britain) during the present century has been matched by a northward shift in the distribution limits of various northern birds (*cf.* Nethersole-Thompson, 1966). There is abundant evidence of the gradual displacement of whole faunas and floras under the influence of climatic changes during the Quaternary Period, and in any one area species at the edge of their range climatically are the first to respond thus as conditions change. According to this theory the climate of the hills of northern England has become marginal and sub-optimal for the dotterel, whereas that of the colder more 'Arctic' hills of the Highlands is still quite suitable. A gradual shift of the migration stream northwards would accord with the decreasing springtime appearances of dotterel in the lowlands of England.

This theory is supported by the indications of incipient recovery of the dotterel in northern England during the last ten years or so, in parallel with the recent climatic reversal towards colder conditions in Britain. There has been an increase in the number of sightings and nestings during the last decade, which has been noted for its run of cold, backward springs. While it is also true that there have been more bird watchers on the hills in recent years, my own observations certainly suggest that this apparent increase is real. The next ten years will be most interesting to watch; a significant recovery would tend to confirm the view that the position of the dotterel in northern England depends largely on climate.

The present status of the dotterel as a summer resident in England is certainly still precarious and the distribution records have to be treated with the utmost secrecy, in order to minimise deliberate disturbance. Any improvement in status must depend not only on favourable climate but also on conservation of the nesting habitat and avoidance of too much interference with nesting birds. Much will depend on the discretion of those fortunate enough to encounter the dotterel on its nesting grounds. For the very elusiveness of the bird can only be a spur to naturalists who find a compelling challenge in the pursuit of difficult subjects. Enthusiasts will continue to search the tops, but all of us who wish the dotterel well must give a thought to the disturbance we ourselves may create.

ACKNOWLEDGMENT

In preparing this account, I have drawn heavily on information freely and generously given me by the late Ernest Blezard from his personal experience of the dotterel as a nesting bird in Lakeland, and from the knowledge he had gleaned through a lifetime's friendship with local naturalists past and present. Moreover his company during searches for this bird on the tops was an additional personal pleasure.

CHAPTER 20

Distribution and Movements

DOTTERELS are now breeding regularly in northern Scotland, Norway, Sweden, Finland and U.S.S.R. They nest irregularly in northern England and Wales, Austrian Alps, Italy, the Sudetenland and in the south and east Carpathians, Romania. Since 1961 they have nested in some numbers on reclaimed polders in Holland, and in 1930 they attempted to nest in Alaska.

Dotterels usually arrive on the south coast between the third week of April and the fourth week in May. A few arrive early. Early dates include 18th February, 1901, Yorkshire; 12th March, 1961, Cairngorms, Inverness; 22nd March, 1853, and 4th April, 1901, Sussex; 22nd March, 1931, Norfolk; and 8th April, 1908, Lancashire. These are, however, isolated records.

In spring dotterels now pass through England in ones or twos or in small trips. Large groups are no longer recorded in the Midlands or in north and north-east England. But stray birds or groups are sometimes noticed when they rest or touch down. In Lancashire dotterels are scarce or irregular spring passage migrants (Oakes, 1953).

In the nineteenth century large flocks, arriving in late April and early May, regularly rested on heaths in Norfolk and Suffolk. A few still continue to do so. One Cambridge heath is still known as 'Dotterel Hall'.

In the 1880s flocks habitually arrived towards the end of April and rested on the north wolds of Lincolnshire and on the Humber marshes, but this no longer happens. In the 1930s, however, small trips continued to visit at least one large farm near Boston in Lincolnshire.

Gunners used to waylay dotterels on the wolds and coasts in Yorkshire; they are now only occasionally recorded on spring passage.

Large spring flocks no longer visit the Solway Marshes in late April and early May. The last sizeable group was recorded in 1928 but dotterels still travel up the east of Lakeland (E. Blezard).

Dotterels seldom visit south Wales lowlands, but in early May they often drop down on Brecon Hills. On 10th May, 1952, a trip rested above Elan Valley, Radnor. In spring dotterels also sometimes rest on

FIG. 14. *World distribution of Dotterel (1945–79)*
Shaded areas show all known breeding grounds. (*Note:* Dotterels nest irregularly in England and Wales, Austria, Italy, Poland, S.E. Europe and Alaska.) Hatched areas show regular wintering areas.

Plynlimmon on the borders of Montgomery and Merioneth. In Caernarvon, the Carneddau are also probably a regular staging post.

Large flocks no longer visit the border counties, but a few still visit the Lowlands. On 16th May, 1967, H. Milne-Redhead watched eleven dotterels in a sandy field and on 16th April, 1965, Donald Watson heard a dotterel call as it flew over Dalry just after midnight.

Dotterels are scarce passage birds on Scottish islands, but they have turned up on Bute, Arran, Islay and Tiree. On 19th May, 1937, two were seen on Mull and in August, 1938, on North Rona and the Flannans.

Orkney and Shetland produce few records; but dotterels occasionally drop down on Fair Isle.

Dotterels are less often recorded on late summer and early autumn passage. On the foggy nights of 29th–31st August, 1914, however, 'hundreds' flew around the Isle of May lantern. On 29th August, 1914, a large flock also appeared at Mull of Galloway Lighthouse. But most records are of small trips from 16th July onwards. The main passage takes place in early August.

In autumn some dotterels appear to go south along the Pennines. On 13th September, 1932, for example, a juvenile hen was picked up dead at Wigton, near Carlisle, and on 3rd October, 1930, J. S. Gayner saw two on Black Combe (1,961 feet), thus suggesting that dotterels return by a more westerly route.

A dotterel occasionally stays late on a Lakeland fell. On 20th October, 1961, A. Watson, Sen., saw a hen on Fairfield, Helvellyn.

In Yorkshire and Lancashire dotterels are fewer in autumn than on spring passage. Even in years of plenty there were never large flocks on the Humber marshes – only ones or twos in autumn. In autumn a few dotterels pass through Wales.

Flocks and trips still occasionally travel through East Anglia and the midlands. On 20th August, 1959, C. D. T. Minton watched a group of forty-seven 'mostly adults, partly in summer plumage' on a Norfolk heath. But recent records mostly talk of solitary birds or small trips.

In late summer a dotterel or two sometimes visit midland sewage farms.

This pattern also obtains on the south coast where in September there are many records of single birds or small groups.

IRELAND

In Ireland dotterels are extremely scarce passage migrants. They have never been known to nest. 'The dotterel has occurred 4 times in April, 3 times each in August, September and November, and once in January, February, March, May, June and October. It has not occurred in July and December.' (P. J. Kennedy et al., 1954.) Dotterels were recorded in Cork, Waterford, Wexford, Tipperary, Down, Antrim, Derry, Donegal, Dublin, Kerry, Offaly, Roscommon, Cavan, and Mayo.

On 14th March, 1948, a trip of about twenty-five was seen just below the summit of Stradbally (2,627 feet), Co. Kerry, and on 26th March two separate trips of about twenty to thirty dotterels were feeding a few hundred yards apart on Slieve Mish at about 1,814 feet.

NORWAY

In Norway dotterels nest on fells above the tree-line from Hekfjeld in Telemark (lat. 59° N) to smaller hills around North Cape. In southern Norway many nest on the great fell systems like Dovre Fjell, Hardanger Vidda, and Nore Fjell. Hardanger Vidda (lat. 60° N) is a stronghold. Dotterels are equally well distributed on the wild fells surrounding the great inlets of the Arctic Ocean – Porsanger, Lax and Varanger fjords about 10° farther north. In Finnmark, however, dotterels choose as nesting fells tops about 300–1,000 feet above sea-level, lower than those in southern Norway.

Earlier accounts of dotterels in Norway omit mention of fells around western fjords. Within recent years, however, Norwegian ornithologists have found nests and broods on the mainland and on at least two large Atlantic islands – Averöya near Kristiansand and Bremangeröy, north of Sognefjord (61° 50'). They have also nested on the Stadtland (62° 15'), a great mainland cape. Further exploration of wild western fjord fells will doubtless lead to many new discoveries (Blair).

Dotterels occasionally nest below the tree-line on fjells in south Norway.

Little is known about movements on the south coast of Norway. Collett and Olsen suggest that dotterels are more frequent spring than autumn visitors. At Lister and Jaederen, however, our birds are scarce but regular visitors in August and September. Nevertheless, in the south of Norway, modern observatories make no mention of dotterels. In Oslo district dotterels are irregular visitors. Occasionally, however, they arrive in good numbers and then frequent upland farms for a few days.

Dotterels usually return to southern Norway in the second half of May and immediately make for highland pastures.

In east Finnmark the earliest dotterels arrive between 22nd May and 15th June. There are few records of the passage of dotterels anywhere along the sea-coasts between Jaederen and Finnmark; but on 2nd June, 1923, George Bolam saw a trip at Bodô, well within the Arctic Circle.

Collett believed that, after breeding in Finnmark, dotterels flew east, while those from the southern highlands crossed the south coast into Denmark and western Europe.

SWEDEN

In Sweden dotterels nest on all suitable fells from the frontiers of Arctic Norway and Finland, into Dalarna (60° N 30'), where pairs or nests have been recorded on Städjan, Himmerasen, Fulufjall, Hus-klappen and Storvätteshågna. In Härjedalen dotterels breed on San-fjället, Groivefjället, Hamrafjället, Rôfjället, and on Flatruet, Skenors-fjället and Rutfjällen and on Mittåfjällen.

In Swedish Lapland dotterels not only nest on great plateaux and tablelands but also on isolated fells, such as Galtispuoda in Arjeplog, in Pite Lappmark and Dundret by Gällivare in Lule Lappmark.

P. Rosenius considers that passing dotterels are seldom seen between departure from Denmark and Germany and their appearance on nesting grounds. They apparently cover the last lap from the Continent to Swedish highlands in one flight. Occasionally – as in 1941 and 1944 – bad weather halts their movement.

Every autumn, usually between 11th and 18th September, dotterels pass through the Idre district in Dalarna and Kvismaren in Narke. They have also been reported in September from Dalsland and from Vadsbo district in Västergottland. There are also autumn records from Rao, Onsala (12th September, 1946) and several from Morupa Tange in Halland (e.g. one bird on 13th September, 1941, seven on 10th September, 1946, and two immature birds on 12th August, 1950).

In other parts of southern and eastern Sweden, migrant dotterels are seldom seen but scattered records include three from Skårne and a few from Oland.

Dotterels return to Sweden in May and June. In May they have been shot at Vargarda, north-east of Alingsås, in Bohuslän, and on the skerries of Göteborg. On 29th May, 1938, six were also seen near Bôden.

Two Swedish-banded dotterels have been recovered. One ringed at Arvidsjaur in Swedish Lapland was reported in Belgium on 8th September, 1946. The other bird ringed on 12th June in Lapland was recovered on 12th November in Dalmatia, Yugoslavia.

In Estonia, Poland and Hungary dotterels appear in small numbers both in May and in August–September. From these scanty records Curry-Lindahl infers that some Swedish-nesting dotterels follow a direct north to south route. From Scandinavia, as a whole, Lindahl suggests that the dotterels generally migrate on a southerly and south-westerly course, presumably making long flights without resting and probably keeping to the highlands. In autumn, however, some Swedes possibly move along the great plateaux into Norway and then fly on to Denmark where they then fly south, south-west, and south-east.

Dotterels probably occasionally winter in Scandinavia.

FINLAND

From Arctic Sweden the dotterel ranges into Finnish Lapland, where it nests on high fells in Enontekio down to Ounastunturi and Pallastunturi.

Dotterels are particularly plentiful on the Ailigas Fells in Utsjoki and on the Värriötunturi Fell. They also nest in the Enare district in Saariselka and as far south as Sallantunturi.

Dotterels are seldom seen on passage in south and central Finland, but probably migrate over the wild wooded districts of east Finland to the Caspian Sea. They are seldom recorded on the well-watched resting places of waders in south-west Finland (Palmgren).

Professor A. I. Ivanov has kindly prepared this detailed account of the dotterel's breeding distribution and movements in U.S.S.R.

(a) The tundra zone from the Kola Peninsula to the Chukchee Peninsula. As the dotterel avoids low-lying swampy tundra there are many gaps in this range.

(b) The alpine zone in the Urals and the mountains of north-east and south Siberia.
The details are:

Kola Peninsula. The tundra zone along the northern coast of the peninsula and the alpine zone in the mountains in the western and central parts of the peninsula: Khibiny Mountains, Chunatundra, Nyavka-tundra and Pel-tundra.

Kanin Peninsula. Not very common in Kanin Range.

Tundra between Kanin Peninsula and Urals. Absent.

Novaya Zemlya. Probably breeds on the southern island.

Kolguev I. and Vaigach I. Was recorded during autumn migration. No records in the breeding season.

Timan tundra and Lower Pechora. Absent (G. P. Dementiev and N. A. Gladkov, 1951).

Yamal Peninsula. Recorded in different places.

Gyda Peninsula (between Ob and Yenesei Rivers). Absent.

Lower Yenesei. Absent.

Yenesei Gulf and adjacent islands. Rather common breeding bird, particularly on Sibiryakov I. (= Kuzkin I.) on shores of Dixon Bay, etc.

Taimyr Peninsula. Known in many places on west, north-west, and east coasts, and in central parts of the peninsula to Taimyr Lake, Yamu-torida River and Boganida River. The most northerly point is Dick Bay (76° 48′ N, 100° 50′ E).

Putorana Mountains. Essei Lake (68° 30′ N, 103° E).

Anabar River. Not very common along the eastern tributaries of the river (middle course).

Lower Lena. Recorded in Balun (approximately 71° 45′ N), south-eastern part of Lena delta and the Kharaulakh Mountains along east side of Lower Lena.

Does not nest in low-lying swampy tundra between Rivers Lena and Kolyma.

New Siberian Archipelago. No definite breeding records.

Chtyrekhstolbovi Island, near Kolyma delta. Recorded in the breeding season.

Chukchee Peninsula. Recorded in Providence Bay and other places.

Anadyr River. Recorded in the suburbs of Anadyr. Very rare as nesting species (Portenko, 1939).

Kamchatka Peninsula and Kuril Islands. No records during the breeding season.

Mountains of N.E. Siberia. Known as a breeding species on Verkhoyanski Range (upper course of Yana River) and the Cherski Range (upper course of Indigirka River).

Mountains of S. Siberia. Mountains near Lake Baikal, Barguzin Range, Baikal Mountains, Khamar-Daban Range, E. Sayan Mountains (Munku-Sardyk Mt. in Tunka Alps), W. Sayan Mountains, Tannu-Ola Range, Altai and Tarbagatai Ranges. Dotterels do not nest in the Tian-Shan and Pamir-Alai mountain systems.

Urals. Recorded in many districts in polar and northern parts of the range. Owing to absence of high mountains in Middle Urals dotterels do not nest there. There are two small separate ranges in the southern Urals. On the highest of these mountains dotterels nest on Iremel Mountain and Yaman-tau Mountain (1,566 metres and 1,638 metres) above sea level.

Winter Quarters. In winter dotterels were recorded as great rarities near Lenkoran (south-west coast of Caspian Sea).

Movements and Migration

No ringed dotterels have been recovered in U.S.S.R.

Spring movements. In the southern parts of Ukraine dotterels have been recorded from the end of February until the end of April (occasionally until end of May). It is possible, however, that dotterels recorded in May were subadult non-breeding birds.

G. P. Dementiev and N. A. Gladkov (1954) give these additional records.

Vil'konskii (1897) reported that dotterels passed through Georgia from early until late April. On 2nd April Lyaister and Sosnin (1942) watched passage birds in Armenia and late in March Senitskii (1898) saw huge flocks in the Crimea. This passage lasted about a fortnight.

Around Kherson dotterels move between mid-March and mid-April. Klimenko (1950) records early arrivals on the Black Sea Reserve and Madame Vladimir (1948) on the Lapland Reserve between 24th May and 5th June.

In 1889 passage was observed at Chkalov from 1st to 7th May. On Emba a movement continued from 8th to 17th May and on the lower Irgiz River from 1st to 2nd May. On 12th May a dotterel on passage was 'obtained' on Vozrozhdenie Island in the Aral Sea and on 27th April two were also taken at Dzhulek.

Lavrov (1930) watched dotterels at Lake Kurgaldzhin in May and on 7th June Bianchi (1902) recorded a movement at Kokchetav.

Shnitnikov (1949) also saw a passage of dotterels in the middle and end of May on Lake Balkhash.

On 15th May a dotterel was seen in the Zaisan Depression. At Novenskoe, in the steppes north-west of the Altai Mountains, dotterels arrived between 9th April and 5th May and at the end of April in Krasnoyarsk.

In early May dotterels reach the Arctic Circle and about 23rd June start to appear on the tundra. On 24th June Torgashev watched passage dotterels at the mouth of the Bolshaya Balakhnya in Taimyr.

At Lake Baikal, Vorob'eva (1931) recorded spring arrivals in the middle or end of May and departure about 6th June.

In Kazakhstan steppes the spring movement lasts till the end of May. Near Krasnoyarsk (Central Siberia) migration takes place at the end of May; near Yakutsk from mid-May till early June; near Magadan (Okhotsk Sea) migration lasts from 20th May till 2nd–5th June. No movements were recorded on Southern Kazakhstan or in Turkestan (Turkmenia, Tadzikistan, Ubekistan and Kirghiz SSR).

Earliest spring arrival records in breeding areas

Kola Peninsula: 26th–28th May (northern parts); 24th May (Khibiny Mountains).
Yamal Peninsula: 1st June (Novyi Port).
Taimyr Peninsula: 7th June (Boganida River).
Putorana Mountains: 1st June (Essei Lake).
Lower Lena River: 4th June (Kharaulakh Mountains).
Kolyma River: 28th May and 7th June (Verkhne-Kolymsk), 4th May (Nizhne-Kolymsk).
Yana River (upper course): 29th May.
Bargusin Range and Chuya Alps (Altai Mountains): 5th and 8th June.

Autumn movements

Most males and young birds leave their breeding grounds in E. Siberian tundras at the end of August, but several dotterels were recorded much later: 7th September (Taimyr Peninsula); 27th September (Lena Delta). The local autumn migration of females begins in the second half of July, but females can be met in the tundra zone until mid-August.

Dementiev and Gladkov give these details.

Middendorff recorded the last dotterels at Boganida on 8th September. Close to Krasnoyarsk first flocks recorded on 2nd September but main passage takes place between 18th and 23rd September. At Novenskoe Village dotterels pass through from mid-August until late September. Flocks were recorded on south shore of Lake Zaisan on 31st August.

Sushkin (1938) did not meet with dotterels after 27th August on their nesting grounds in the Altais, but they apparently did not leave breeding-biotopes in the western Sayans until 10th December or later.

In N.E. Kazakhstan movements take place from late August to late September. In Semireche only one dotterel 'obtained' (7th October, Tentek Valley). A dotterel was also taken near Kazalinsk at the end of September and another at Uzun-ada in S.W. Turkmenia about the same date. On 27th September a large flock was recorded on Emba and dotterels are on passage near Chkalov from late August to late September. N. A. Severtsov also recorded a movement on 20th–22nd October at Terekli Observatory. Passage through west Siberia is irregular – often only stray birds. Main movements seem to pass close to eastern boundaries of west Siberia in the Altai steppes, in the Zaisan depression, along the west banks of River Emba and western slopes of Mugodzhary Mountains, thence probably into lower reaches of Ural to Gurev.

There are few records about movements in European U.S.S.R. Latest dotterels recorded on 26th August in Lapland Reserve and in August and September round Kazan. Several recorded Minsk 22nd September. Infrequent in Poltava region, 12th September–11th November. Passage through Black Sea Reserve takes place from late August until early October. Near Mariupol considerable movements recorded in mid and late September. Quite large numbers sometimes pass through Ciscaucasia in October and November. Main passage in Transcaucasia is from early October to second half of November, but a few dotterels still stay as late as December. Zarudnyi recorded three dotterels on 19th September between Hamun and Baud-i-Zirreh.

During their passage-movements dotterels do not follow rivers and valleys but usually fly straight across the steppes. During their main movements the dotterels often travel in groups of five to fifty birds.

They take off at daybreak and after flying low over small hills they drop down in the lowlands. They then feed from about noon until 4 p.m. and then fly on but come down before nightfall (Sushkin, 1908).

The latest records of the dotterel in U.S.S.R. are: 13th September (Kirensk, Upper Lena): 18th September (Baikit, Podkamennaya, Tunguska River); 16th September (Pavlodar, Irtysh River); 22nd September (City of Minsk); 7th November (Makhach-Kala, north-west coast of Caspian Sea). These were all young birds.

DENMARK

Dotterels do not nest in Denmark, but are extremely abundant on passage where they follow different routes on their spring and autumn movements (Salomonsen). Spring passage extends from late April until late May and most dotterels then travel up the west coast of Jutland. In autumn, on the contrary, most travel down the east of Jutland from the end of August to mid-October.

CONTINENT OF EUROPE

In western Europe, apart from Britain, dotterels nest only in Holland, where fifty or more pairs breed. The dotterel was previously only known in the Netherlands as a passage migrant, usually on the coast, exceptionally inland. Spring passage lasts from late April until June. Early dates include early April, 1942, Deventer, and 9th April, 1950, in east polder. Before the dotterels started to nest, large groups and trips sometimes passed through Holland.

From mid-August until late September dotterels appear in smaller numbers. The province of Limburg is a favourite resting-place. Trips also stay for a few days at the end of August and in early September in the Beer Sanctuary (N. Ruitenberg).

On the Continent of Europe dotterels continue to nest periodically or irregularly on high alpine grasslands in Austria, Poland, Italy and Romania (R. Heyder, 1960-2).

AUSTRIA

In Austria dotterels formerly nested in the Zirbitzkögel and Wenzel Alps, west of Judenberg, and the Sau Alps, farther south. But the late nineteenth century produced few acceptable breeding records.

Nearly a century later dotterels have again nested in the Austrian Alps. In 1947 Hans Franke heard of their return to the Seetal Alps and on 24th July, 1948, he saw one dotterel on a high alpine meadow. In

1949 Franke found four nests in high alpine meadows and in 1952 three nests. In 1951 and 1952 Father B. M. Stenger also found dotterels nesting on the Carinthian slopes of the eastern alps. These scattered records do not suggest that the dotterel is a genuine relict species in Austria.

SUDETENLAND

Dotterels nest irregularly in the Sudetenland, now largely Polish territory, on the Czechoslovak borders. Between 1825 and 1948 dotterels were only recorded in twenty-five years. Nests or broods, however, were found on the Riesengebirge in twelve years, 1825, 1826, 1846, 1858, 1865, 1870, 1874, 1877, 1882, 1887, 1903 and 1946 (R. Heyder, 1962). About one hundred years ago dotterels sometimes nested in high numbers on some of these mountains. In 1858 Anton Fierling, a dotterel-hunting chemist, alone 'procured' seventeen dotterels and took thirty eggs.

There are few acceptable records for the twentieth century. In 1903 Johann Bonsch found a pair with chicks and in July, 1922, Martin Schlott saw a dotterel on Prinz Heinrich Baude, north-west of Schneekoppe. In July, 1937, Breslau ornithologists watched two dotterels; one allegedly with 'food in its beak'.

On 28th June, 1940, Heinz Krampitz met with a pair and on 30th June, 1946, Dr Josef Maran found a cock and brood.

Several observers have recorded dotterels in summer on the Isergebirge. In late nineteenth century dotterels also possibly nested on the Altvater hills.

SOUTH AND EAST CARPATHIANS

Dotterels sometimes breed in the south and east Carpathians on the borders of Hungary and Romania. In 1869 a cock and two downy young were seen on Csindrel in the south Carpathians; and in 1895 M. von Kimakovicz saw two young and a trip of twenty-three adults south of Sibiu, Romania. Then, in May, 1904, V. Metz found a nest on another hill. There are no more acceptable records until 1960 when Professor W. Klemm's son found a nest and in late July photographed large unfledged chicks. In early July, 1961, he also found an addled egg and recently hatched chicks.

In 1958 J. von Beress watched a dotterel giving distraction displays on Radnaer Hills, Rayon Seghet. A second dotterel flew overhead. Dotterels have thus returned to the Carpathians during this cooler climatic phase.

ITALY

Dotterels occasionally nest in Italy. In early June, 1939, Camill Gugg watched one on Monte Marsicano, north of River Sangro. Then on 19th July, 1952, R. Vaughan saw a pair and chick on Monte Amaro in the Abruzzi. In June, 1953, he again watched two dotterels there.

GREECE

Between 17th and 21st June, 1956, Fritz Peus heard dotterels calling on Olympus in Thessaly, Greece – a possible new European nesting haunt.

Recent extensions of range

In the last twenty-five years dotterels have gained new, or re-claimed old, territories south of their normal range. They started breeding in Holland (1961 onwards), returned to English fells (1956, 1959, 1960, 1968–71), Austrian Alps (1949, 1951, 1952), Riesengebirge, Poland (1946) and south Caucasus (1960 and 1961).

ICELAND

Dotterels have probably bred or attempted to breed in two other countries. In the British Museum Collection I found two eggs marked 'Iceland, 2 June 1864', but could discover no precise details.

U.S.A.

Dotterels have occasionally crossed the Chuckchee Sea from U.S.S.R. to Alaska, U.S.A. On 23rd July, 1897, one was killed on King Island. Between 15th and 19th June, 1929, an Eskimo shot two at Wales, Alaska. Then, on 14th June, 1930, an Eskimo boy shot a cock dotterel near Barrow. Two days later he killed the hen with a complete egg in her oviduct. 'This was the first and so far the only breeding dotterel recorded in Alaska, but another hen was shot on 6 June 1931. In May and June 1931 two dotterels of unknown sex, but in fairly worn plumage, were also procured at Gambell in the north-west corner of the St. Laurence River.' (H. Friedmann, 1932.)

Since 1931 American ornithologists have found no more dotterels in Alaska.

Passage

Small trips pass through northern France on spring and autumn passage. A few formerly wintered in the Alpes Orientales.

In Belgium dotterels are regular passage migrants in reduced numbers. Most spring dotterels are recorded at the end of April or early in May. The autumn movement is in August and September, exceptionally later. The dotterels usually favour the western lowlands and midlands, seldom visiting north-east Belgium, but they are occasionally recorded in La Campine and exceptionally at Ossendrech, north-east of Antwerp. On 27th September, 1927, dotterels were also recorded at Hingeon in the Ardennes.

Scandinavian dotterels formerly migrated through Germany, but this old route is apparently no longer used. Sporadic records now usually refer to single dotterels seen near the German coast. In October, 1959, however, Professor F. Steiniger met with two adults and six young at Hildesheim.

In Switzerland dotterels are scarce. The occasional records are almost always of autumn migrants.

Dotterels sometimes pass through Austria, but some records possibly refer to birds bred in the Alps. On 31st August, 1952, for example, Franke watched eleven flying south over the mountain tops but it was impossible to determine whether these were nesting or passage birds. The birth-place of dotterels seen in central and south-east Europe on spring passage is likewise obscure.

Dotterels pass through Yugoslavia. I have already referred to the Swedish juvenile recovered in Dalmatia.

In the western Mediterranean dotterels are usually scarce on spring passage. In Malta dotterels are scarce in spring, but were formerly in plenty on autumn passage (early date 23rd August, late date 11th December). Dotterels pass through Sicily and formerly regularly visited Sardinia, where some periodically spent the winter.

In Greece dotterels are scarce passage migrants and winter visitors from October to March. On 10th September, 1959, Ballance and Lee recorded one in the Axios Delta. For Cyprus there are apparently no spring records, and only three 'sightings' in winter.

Great tundras in U.S.S.R. carry most of the world's breeding populations of dotterel. In August and September, therefore, the great mass of migrants pass through the Soviet Union into eastern Europe before flying on to Israel, Palestine, Syria and the North African countries where every year most dotterels winter.

R. Meinertzhagen (1921) made two measurements – 45 mph and $50\frac{1}{2}$ mph – of the ground-speed of autumn trips in S. Palestine.

Winter range

C. Vaurie (1965) defines the dotterel's wintering range as north-west Africa (Hauts Plateaux) and southern Tunisia and the basin of the Mediterranean – particularly in the east Mediterranean and eastwards to Iraq and the Persian Gulf. Dotterels also winter in smaller numbers beside the Caspian Sea.

R. Meinertzhagen (1954) mentions that dotterels are common winter visitors to Palestine, Iraq and Sinai and occasionally in Egypt, but that they do not reach the Arabian Peninsula although often occurring in the Syrian desert. In the first week of November dotterels arrive in north Sinai in trips of ten to twenty and stay there in high numbers from early December onwards. In January they start to move on and all are gone by March. In winter they live in fairly compact flocks. Dotterels are regular winter visitors to Tripolitania from about 22nd October to 17th January (K. M. Guichard, 1956) and they arrive in Tunisia in September and October and go away in March.

In Morocco K. D. Smith (1965) recorded 150 in his first, and 175 in his second winter there. On 27th and 28th September, 1963, the earliest to arrive was a party of twenty-one on the coast near Oualida. Between mid-November and mid-February there were flocks between the Tiznit plains, right across the steppes north of the High Atlas near Fkih-Bensalah, and on the plains near Berguent. A few dotterels favoured the Middle Atlas moorlands near El-Hajeb.

In the nineteenth century Canon Tristram remarked on 'the continuous flocks which overspread the whole of the southern wilderness during a three-days ride from Arabeh to Beersheba. Hour after hour the birds ran almost among our horses' feet and we shot as many as required for the day's provision within half an hour.'

In 1851 von Huglin also saw a large flock on the desert between Sahara and the Fayoom and met with trips along the western shores of the Red Sea.

Dotterels do not reach the interior of Egypt or the Sudan, but Tristram referred to vast flocks in the Sahara. Reports that dotterels winter on the Blue Nile are possibly due to confusion with caspian plovers (D. A. Bannerman, 1961).

Dotterels sometimes stray to strange exotic places like the Canaries, Madeira, Bermuda, Faroes, Bering Island, Kuriles, Sakatang and Japan as well as north-west United States (Vaurie).

CHAPTER 21

The Future

WE now know more about dotterels than when I first went to the Cairngorms and Grampians in 1933. We can answer all Jourdain's basic questions. But there is still much to learn.

Living in well-equipped huts on the mountain tops, teams of dotterel watchers would discover much about arrival and dispersal patterns. They could undertake round-the-clock watches which are out for those who only have small tents. Colour-ringing of adults and chicks might also clarify many problems but recent experience in Finland and Sweden has hardly been encouraging.

In a society under Petticoat Government are there really a few high status hens (possibly like Blackie) which often mate with two or three cocks and perhaps produce a high proportion of flying young? Or did Blackwood almost discover a still unknown aspect of polyandry, with perhaps two cocks sharing a hen and then both brooding her eggs in the same nest? Is there any substance in my guess that many members of the large nineteenth-century flocks were destined to become non-breeding surplus birds? In the future, close observation of non-breeding groups in Holland is an urgent task.

We now know more about brooding-behaviour, but why do a few hens incubate, but most apparently do not? And, why is this adaptation less decisive than in phalaropes? We have measured incubation periods in Scotland and other countries, but why do these so greatly vary? It is easy to explain the longer periods in terms of individually differing brooding rhythms and eggs left to cool, but more difficult to understand why some measurements are so short. We need many experiments with incubators before reaching firm conclusions.

Have we really proved that some hill groups carry nesting populations which produce a significantly higher ratio of flying young to adults than those breeding on the granite barrens? Is this due to a greater wealth of insects and do these young dotterels mature more quickly than those elsewhere? Does the crop of young fledged on these hills ultimately create a floating surplus containing prospective colonists and pioneers? Do young dotterels usually breed on their first summer and do they often return to their native hills? Or have dotterels, as we now believe, evolved a weak *ortstreue*? Why do dotterel chicks often die in their first few days? Is this through lack of food, rough weather, single-parent patterns, or to more fundamental causes? I have made suggestions which others can prove or disprove.

We now know more about elements in the dotterel's world, but there are still many ambiguities to reconcile. I once believed that a hill – however small – was essential. Think of those small rounded hillocks and tongues of fell dispersed on the immense rolling tundras of the Yenesei. But dotterels have now nested successfully on new polders in Holland which are actually below sea-level. Are they now merely using the living-space from which better adapted competitors ousted them in distant inter-glacial phases? Did the competition of stone curlews, for example, formerly prevent dotterels exploiting suitable ecological niches on downlands in Sussex and Wiltshire or on the stony brecks and commons of East Anglia which they still often visit on spring passage? Large non-breeding groups began by exploring Dutch polders on which they later nested.

Do quite small oscillations of climate radically affect the distribution of birds which are breeding on the fringes of their range? Recent experience of dotterels and other boreal birds do seem to suggest this. But what are the triggers?

We still have to discover much more about what dotterels eat in summer, but this is likely to involve much shooting and analysing.

I knew most of the twentieth century dotterel hunters. Their knowledge of our bird's behaviour was small compared with what they had learnt about other hunters' birds, like greenshank and hobby, siskin, corn bunting and woodlark. But I doubt whether their 'clutching' greatly harmed dotterels in Scotland.

Egg-hunting, as a romantic contest between man and bird, is already an obsolescent field sport. The drugged dart, the mist net and the midget radio transmitter are replacing drill and blowpipe. But these new challenges will lead to equally exciting discoveries.

Our greatest boreal ornithologist, Hugh Blair, wrote rather wistfully: 'I feel that the dotterel has rather let the side down. We can no longer think of him running over the misty fells but sitting in the middle of some Dutchman's potato patch.' I feel rather differently. The defeated has risen. For sheer romance, I know of little to compare with the courage of this pioneering. Young dotterel-watchers should go to Holland. There is history to be made on those potato patches and beetroot fields.

We always scorn the follies of our fathers and are quite blind to our own. New pressures are likely to become more deadly to the dotterels than the worst that the old eggers and trophy hunters ever inflicted on them. Toxic sprays are already killing dotterels and their chicks among the flax on Dutch polders. We do not yet know whether their normal winter environments are also contaminated.

Other dangers are more obvious. Skiers covet those same few broad-backed hills and tablelands that the dotterels always seem to need in Scotland. How fantastic that in so large a wilderness man and dotterel are in such strange and deadly competition. Developers have already built a motorway high up Cairngorm, where chairlifts carry thousands to the edge of our greatest Arctic relict. They plan another road through Lurcher's Gully on Creag an Leth Choin. In the west Cairngorms an estate motor track already winds up to the top of Carn Ban Mor, where Thornton once flew his hawks. In the Drumochter Hills they speak about spur roads to ski runs on Marconnaich and Sgairneach Mor.

Deeside Planners urge a road up Beinn a' Bhuird and suggest another to connect it with Ben A'an. Herr Panchaud of Switzerland has already bulldozed a spectacular Land-Rover road almost to the top of Beinn a' Bhuird. There is also a ski-tow on Glas Maol in Angus.

The Highlands badly need industry and tourism. Jobs for our young people are scarce. But development will create new problems. These we must accept and solve.

Industry has already reached Easter Ross. North of the Great Glen skiers and businessmen have surveyed the slopes of Ben Wyvis where they hope soon to build a motorway, spur roads, chairlift and cafés. Beinn Dearg and the wild country around will soon be at risk. This is only the beginning. We have to create good winter and summer playgrounds without destroying the finest samples of our hill birds' heartlands.

Here and overseas Conservationists deliver weighty speeches at international conferences and innumerable committees meet, discuss, and talk. At no other time have so many wished so well of our Highland wilderness. But we do need vision and action. We are still waiting for a balanced plan reconciling conflicting interests.

As a boy I read about the Highlands and their birds. And when I came here they were just as I had seen them in my dreams. You may find them changed. Until recently the line of Highland roads had altered little since Booth and the late Victorian trophy hunters had explored the north by gig and trap. Old tracks were merely tarmacadamised to adapt them for the motor-car, but they still followed the valley, leaving the remoter mountain tops to the red deer, hill birds, and walkers. Now, without any coherent plan, the developers are moving in. Tracks, roads, spur roads, chairlifts, hovercraft, and possibly vertical landing aircraft, in that order, and finally the hordes spilling over and polluting every whaleback. The Highland weather is soon likely to be the only factor favouring the dotterels.

Vernacular Names

Dutch Morinelplevier, morinel.

English Dotterel, Dotrel, Dottrel, Dotteral, Ash dotterel (Lancs.), Daft dotterel (South of Scotland), Mical Dotrel (Old Naturalists).

Finnish Keräkurmitsan.

French Pluvier Guignard.

Irish Amadan Mointich.

German Mornellregenpfeifer, Mornell.

Italian Piviere tortolino.

Scottish (Gaelic): Amadan Mointeach=The Foolish Fellow of the Peat Mosses or Mossfool.

Swedish Fjällpipare.

Russian Khrustan, also Glupaya Sivka=The Stupid Steed.

Welsh Hutan.

Description and Moults

Ernest Blezard describes a chick about a week old, picked up dead on a Lakeland fell on 29th June, 1968.

'The upper parts, including the crown of the head, mottled brownish-black and golden brown. Touches of white about the rump. The whiteness of the forehead, running back over the eye, divided by a dark sepia line up from the base of the beak to the crown. This white patch bordered above the eye by a dark sepia line shortly meeting another encircling the crown to enclose, behind the eye, a beautiful oval of warm apricot colour. A spreading white patch at the back of the crown above the longer line and, below this line, a broad white band around the nape to the chin, clearly separating the design of the crown from that of the body and much more clear than the nape patch of a peewit chick, whether in a "disruptive pattern" or not. The near white underparts, with dusky mottlings, showing the young breast feathers, just beyond the down, in a richer, warm apricot colour.

'Beak grey-black; legs and feet beechbark-grey tinged with a very pale and delicate green; iris dark brown.'

Between March and June dotterels gradually moult from winter to summer dress, hens probably attaining full nuptial plumage ahead of cocks. Some dotterels thus occasionally reach their nesting grounds in Britain before the moult is complete. They then lack the brilliant chestnut-red lower breast and flanks and the deep black belly patches, but have dull buff eye-stripes, grey brown mantles, pale brown lower breasts and flanks and indistinct off-white breast-bands. In the second half of July this is reversed. The winter moult now begins and the dotterels gradually lose their summer brilliance. We often watched this in the Cairngorms.

In U.S.S.R. the moult to winter plumage largely occurs on migration. In early September the moult of small feathers is about half complete and by the end of October observers recorded full winter plumage in Kazakhstan. Juveniles partially moult from September to October, but a few winter feathers show on the mantle at the end of August. (G. P. Dementiev et al., 1951.)

Fledged young have indistinct eye-stripes and breast bands. Heavy blackish marks on back and wings and their generally creamy ground colour are good fieldmarks. (A. Watson.)

Witherby (1940) adds that 'the juvenile body-plumage (not all the scapulars nor all the feathers of back and rump) sometimes the central pair of tail-feathers, some innermost secondaries and coverts and some median and lesser coverts, are moulted from Sept. to Nov., but not the rest of the tail feathers or wings.'

Scottish Deer Forests in which Dotterels have been seen in summer

(*Note:* Nesting has not been proved in all.)

Forest	Highest Hill	Height (feet)
Aberdeen (4)		
Balmoral	Lochnagar	3,786
Glencallater	Glas Maol	3,502
Invercauld	Beinn a'Bhuird	3,924
Mar	Ben Mac Dhui	4,296
Angus (2)		
Caenlochan	Glas Maol	3,502
Glendoll	Tam Buidhe	3,140
Banff (1)		
Inchrory	Ben Mac Dhui	4,296
Inverness (21)		
Aberarder	Carn Liath	3,298
Abernethy	Cairngorm	4,084
Ardverikie	Aonach Beag	3,688
Ben Alder	Ben Alder	3,757
Braeroy	Creag Meagaidh	3,700
Braulen	Sgurr an Lapaich	3,773
Ceannacroc	Sgurr nan Conbhairean	3,634
Coignafearn	Carn Mairg	3,087
Corryarick	Carn Liath	3,298
Culachy	Corrieyairach	2,922
Gaick	Meall Chuaich	3,120
Glencannich	Sgurr na Lapaich	3,773
Glendoe	Corriezairach	2,922
Glenfeshie	Brae Riach	3,950
Glenmore	Cairngorm	4,084
Insch Riach	Sgoran Dubh Mor	3,635
Invereshie	Sgoran Dubh Mor	3,658
Killiechonate	Ben Nevis	4,406
Moy	Creag Meagaidh	3,700
Newtonmore	Carn Dearg	3,093

Rothiemurchus	Brae Riach	4,248

Perth (11)

Chesthill	Carn Mairg	3,419
Craiganour	Sgairneach Mor	3,160
Dall and Upper Finnart	Carngorm	3,370
Dalmunzie	Glas Thulachan	3,445
Dalnacardoch and Sron Phadruig	Carn na Caim	3,087
Dalnamein	A'Chaoirnich	2,841
Dalnaspidal	Beinn Udlamain	3,306
Fealar	Glas Thulachan	3,445
Glen Lyon	Carn Mairg	3,419
Meggernie	Ben Lawers	3,984
Rhiedorrach	Glas Thulachan	3,445

Ross and Cromarty (11)

Beinn Eighe	Ruadh Stac Mor	3,309
Benula	Carn Eige	3,877
Brae More	Beinn Dearg	3,547
Fannich	Sgurr Mor	3,637
Garbat	Ben Wyvis	3,429
Inverlael	Beinn Dearg	3,547
Kildermorie	Carn Chuinneag	2,749
Monar	Sgurr a Chaoruinn	3,452
Strathconon	Sgorr a Choir-Ghlais	3,544
Strathvaich	Cona Mheall	3,200
Wyvis	Ben Wyvis	3,429

Sutherland (2)

Ben Loyal	Ben Loyal	2,504
Kinloch	Ben Hope	3,040

Nests of the Dotterel in Northern England

1784	29 June	3 eggs	Skiddaw, Cumberland	*per* Dr J. Heysham
1835	29 June	3 eggs (12–14 days incub.)	Whiteside, Helvellyn Range, Lake District	J. Cooper
1835	5 July	2 eggs (fresh)	Robinson, Cumberland	J. Cooper and T. C. Heysham
1835	5 July	1 young (a few days old)	Robinson, Cumberland	J. Cooper and T. C. Heysham
c 1838	?	eggs	Robinson, Cumberland	W. Hewitson
1841	7 June	3 eggs	Melmerby Fell, Crossfell Range, Cumberland	B. Greenwell
1842	12 Aug.	1 young	Red Pike, Cumberland	J. W. Harris
1845	28 May	2 eggs	Saddleback, Cumberland	W. Bowe
1845	4 June	1 egg	Helvellyn, Lake District	W. Bowe
1846	12 Aug.	2 eggs	Helvellyn, Lake District	Dickinson
1848	June	1 egg	Red Pike, Cumberland	C. Wilkinson
1849	?	3 eggs	?Cumberland	*per* C. Wilkinson
1851	21 June	1 egg	Skiddaw, Cumberland	*per* Lord Lilford
1852	21 June	3 eggs	Skiddaw, Cumberland	W. Greenup
1853	1 June	3 eggs	Skiddaw, Cumberland	W. Greenup
1853	5 June	3 eggs	Grasmoor, Cumberland	W. Greenup
1854	5 June	3 eggs	Skiddaw, Cumberland	W. Greenup
1854	15 June	2 eggs	Skiddaw, Cumberland	W. Greenup
1855	21 May	1 egg	Skiddaw, Cumberland	*in* Hancock Collection
1855	11 June	3 eggs	Skiddaw, Cumberland	W. Greenup
1855	17 June	1 egg	Helvellyn, Lake District	W. Greenup
1857	23 May	1 egg	Skiddaw, Cumberland	W. Greenup
1857	6 June	3 eggs	Skiddaw, Cumberland	W. Greenup

1857	12 June	3 eggs	Skiddaw, Cumberland	W. Greenup
1860	?	1 egg (incub.)	Cumberland	F. Nicholson
1864	5 June	1 egg	Cumberland	F. Nicholson
1874	29 May	3 eggs	Robinson, Cumberland	W. Greenup
1874	July	3 eggs	Cumberland	F. Nicholson
1874	July	3 eggs	Cumberland	F. Nicholson
1879	25 June	2 eggs	Cockley Fell, Westmorland	J. Watson
1880	?	Brood	Hartside Matterdale	W. Hodgson
1881	17 June	1 egg	Cockley Fell, Westmorland	J. Watson
1883	30 June	3 eggs	Fairfield, Westmorland	J. Watson
1884	1 July	3 young	Grasmoor, Cumberland	H. E. Rawson
1888	?	3 nests	Cumberland	J. Watson
1889	19 June	1 egg extant	Buttermere Fells, Cumberland	Royal Scottish Museum Collection
1889	?	3 nests	Cumberland	H. E. Rawson
late 1880s		1 egg	Crossfell, Cumberland	R. Raine
1894	19 July	2 eggs	Lake District	J. H. Salter
1894	19 July	young	Lake District	J. H. Salter
1902	29 May	1 egg (incub.)	Derwent Fells, Cumberland	F. Nicholson
1902	?	Brood	Yorkshire	R. Chislett
1905	17 June	1 egg (hard set)	Buttermere Fells, Cumberland	J. F. Peters and F. Nicholson
1908	2 June	2 eggs (incub.)	Buttermere Fells, Cumberland	J. Baldwin-Young
1908	7 June	2 eggs (3 later)	Buttermere Fells, Cumberland	E. S. Steward
1910	3 July	young (1 week old)	Buttermere Fells, Cumberland	J. F. Peters
1910	3 July	young (2 weeks old)	Buttermere Fells, Cumberland	J. F. Peters
1911	27 May	2 eggs (slightly incub.)	Buttermere Fells, Cumberland	J. Baldwin-Young
1911	28 May	2 eggs (hard set)	Buttermere Fells, Cumberland	J. F. Peters
1911	4 June	2 eggs (hard set)	Buttermere Fells, Cumberland	J. F. Peters
1911	5 June	2 eggs (hard set)	Skiddaw Group, Cumberland	J. B. Wheat

1911	26 July	young (3–4 days old)	Buttermere Fells, Cumberland	J. F. Peters
1911	?23 May	3 eggs	Cheviots, Northumberland	M. Logan Home
1911	11 June	3 eggs (almost hatched)	Helvellyn Range, Lake District	J. F. Peters
pre-1912		nest with eggs	Durham	*per* H. M. S. Blair
1912	30 May	3 eggs (fresh)	Buttermere Fells, Cumberland	J. F. Peters and G. H. Lings
1912	?	?	Cheviots, Northumberland	G. Bolam
1913	16 June	young (2 days old)	Cumberland	G. Bolam
pre-1914		nest	Cumberland	W. F. Parker
1914	4 July	3 eggs	Cumberland	G. Bolam
1914	4 July	young (very small)	Cumberland	G. Bolam
1915	27 July	1 young (fully fledged)	Cumberland	G. Bolam
1916	8 Aug.	2 young (well fledged)	Cumberland	G. Bolam
1917	17 June	3 young (day or so old)	Cumberland	G. Bolam
1917	28 June	3 birds (and sucked egg)	Helvellyn Range, Lake District	G. W. Temperley
1918	30 May	1 egg	Helvellyn Range, Lake District	J. F. Peters
1918	2 June	2 eggs (fresh)	Buttermere Fells, Cumberland	J. Baldwin-Young
1918	4 June	1 egg	Helvellyn Range, Lake District	J. Baldwin-Young
1919	21 June	2 young	?Buttermere Fells, Lake District	C. Oldham
1920	26 May	2 eggs (rather set)	Helvellyn Range, Lake District	J. F. Peters
1920	27 May	2 eggs	Buttermere Fells	J. F. Peters
1921	25 May	1 egg	Helvellyn Range, Lake District	J. F. Peters
1921	28 May	3 eggs (a good deal incub.)	Cumberland	G. Bolam
1921	29 May	3 eggs	Westmorland	J. Henry

1922	27 May	3 eggs (fresh)	Helvellyn Range, Lake District	J. F. Peters
1922	21 July	small young	Cumberland	G. Bolam
1923	?	nest	Cumberland	J. O. Wilson
1925	7 June	3 eggs (incub. about a week)	Westmorland	E. Blezard
1925	14 June	3 eggs (incub. about 2 weeks)	Westmorland	E. Blezard
1925	28 June	3 eggs (deserted, two broken)	Cumberland	E. Blezard and R. Graham
1925	?	3 eggs	Buttermere Fells	W. H. Pearsall
1925	?	2–3 young (unable to fly)	Buttermere Fells, Cumberland	W. H. Pearsall
1926	31 May	2 eggs (incub. 3–4 days)	Westmorland	E. Blezard
1926	8 June	3 eggs	Cumberland	J. Markham
1932	?	3 eggs	Buttermere Fells, Cumberland	E. Telford
1934	31 May	3 eggs	Cumberland	R. H. Brown
1937	?	2 eggs	Buttermere Fells, Cumberland	W. H. Pearsall
c 1937	?	eggs	Cumberland	J. Irvine
1942–4	?	2 nests	Helvellyn Range, Cumberland	E. S. Steward
1956	8 June	3 eggs	'many miles from where I found my nest,' 2,500 feet above sea-level	per R. H. Brown
1959	31 May	3 eggs (fresh)	Lakeland	E. Blezard and D. A. Ratcliffe
1960	28 May	3 eggs	Lakeland	D. A. Ratcliffe
1968	30 May	3 eggs	Lakeland	D. A. Ratcliffe
1969	25 May	3 eggs	Lakeland	R. Laidler
1969	9 June	3 eggs	Lakeland	D. A. Ratcliffe
1970	17 May	3 eggs	Lakeland	R. Stokoe
1970	5 June	3 eggs	Lakeland	G. Horne
1970	21 June	3 small chicks (possibly from above)	Lakeland	per H. M. S. Blair
1971	late June	3 eggs	Lakeland	per G. Horne

Additional records of dotterel seen on known or possible nesting grounds in spring and summer

1908	3 June	a pair	Buttermere Fells, Cumberland	J. Baldwin-Young
c 1913	3 June	1 dotterel	Buttermere Fells, Cumberland	W. H. Pearsall
1928	28 May	2 birds	Cumberland	R. H. Brown
1931	end of May	1 bird	Crossfell	G. Bolam
		1 bird	Coalcleugh	

Between 1939 and 1945 nest said to have been found by gamekeeper in Westmorland *per* R. W. Robson

1945	July	pair	Glaramara	W. H. Pearsall
1954	May	2 birds	Cumberland	B. Campbell
1954	23 May	1 bird	Cumberland	D. A. Ratcliffe
1961		1 bird	Cumberland–Northumberland border	F. H. Day, jun.
1964	mid-June	1 bird	Westmorland	Anon.
1964	28 May	2 birds	Lakeland	E. Blezard and D. A. Ratcliffe
1967	4 June	1 bird	Yorkshire	Anon. *per* D. A. Ratcliffe
1970	31 May	1 bird	Lakeland	D. A. Ratcliffe
1970	3 June	1 bird	Yorkshire	D. A. Ratcliffe

APPENDIX 5

Parasites

Dr Theresa Clay kindly tells me that the following Mallophaga or feather-lice have been recorded on dotterels. *Actornithophilus ochraceus* (Nitzsch, 1818). Family Menoponidae *Saemundssonia sp. S. semivittata* (Giebel, 1874) has been described from the dotterel but existing material of *Saemundssonia* from this host is insufficient to determine whether this is a good species or not. Family Philopteridae. *Quadraceps punctifer* (Hopkins, 1949). Family Philopteridae.

These parasites live on the host's feathers, feeding on blood and tissue fluids. In size and shape they are adapted to different ecological niches on the bird's body.

Mr S. Prudhoe, of the British Museum, tells me that these parasitic worms have been located in dotterels: Roundworms, (*Nematoda*): *Porrocaecum ensicaudatum* and *P. heteroura*. Thorny-headed worms, (*Acanthocephala*): *Centrorhynchus lancea*. Tapeworms (*Cestoda*): *Anomotaenia microphallos*.

Dr H. H. Williams summarised what is known about these worms. *P. ensicaudatum*, a nematode or roundworm, is particularly common in passerine birds. N. L. Levin (1956, 1957 and 1961) found this worm in American robin, starling and bronzed grackle. The infective larva, located in the ventral blood vessel and heart of *Lumbricus terrestris* and other earthworms, was exsheathed within forty-eight hours between the horny layer and muscular wall of the gizzard. By the third day it was located in the duodenal wall. The second moult was completed by the fourteenth day. On the eighteenth day the worm began to emerge from the intestinal wall into the lumen. The worm then had characteristic features, but continued to grow.

Dotterels presumably become infected by eating earthworms containing the parasite's larva.

A. Mosgovoi and L. Bishaeva (1959) discovered that the infective larvae of *P. heteroura*, another nematode, live in the ventral blood vessels or hearts of earthworms *Lumbricus terrestris*. This nematode also therefore passes into the dotterel when it eats infected earthworms.

The life history of *C. lancea* is still unknown. In *Systema Helminthum*, however, Yamaguti describes the life-cycle of this genus. 'Adults parasitic in birds, especially *Raptores* and mammals; larvae in invertebrates, amphibians, reptiles or mammals.'

There is no recorded life-history of the tapeworm *A. microphallos*, but Soulsby describes the genus as large and occurring in plovers *Charadrii* and gulls *Lari* as well as in passerines.

No flukes *Trematoda* appear to have been recorded in dotterels. Landsnails are, however, the intermediate hosts of flukes. Do dotterel thus risk infection from this source?

Selected Bibliography

ARMSTRONG, E. A. (1964). 'Polyandry' in *New Dictionary of Birds*.

BACKHOUSE, J. (1896). *Upper Teesdale*.

BAKKER, G. (1961). Morinelplevier broedt in N.O. polder, *Het Vogeljaar* 9: 185–7.

BANNERMAN, D. A. (1953–63). *The Birds of the British Isles*, 12 vols., Edinburgh, Oliver & Boyd (Dotterel 10: 237–53).

BARTH, E. (1961). (*The Tameness of Some Scandinavian Waders*), *Brit. B.* 54: 133–6.

BAXTER, E. V. and RINTOUL, L. J. (1953). *The Birds of Scotland*, 2 vols. (Dotterel 2: 595–7).

BEEBE, W. (1925). The Variegated Tinamou . . . , *Zoologica* 6: 195–227.

BENT, A. C. (1929). Life Histories of North American Shore Birds . . . (Part 2), Washington, *U.S. Nat. Mus. Bull.* 140 (Dotterel: 150–3).

BERG, B. (1933). *Mein Freund der Regenpfeifer*, Berlin, Reimer.

BLACKWOOD, C. G. (1920). Notes on the Breeding Habits of the Dotterel (*Eudromias morinellus*) in Scotland, *Scot. Nat.* 98: 185–94.

BLAIR, H. M. S. (1961). 'The Dotterel in Scandinavia' *in* Bannerman, D. A. 10: 242–4.

BLEZARD, E. (1926). Breeding of the Dotterel in the Pennines in 1925, *Brit. B.* 20: 17–19. (1958) Lakeland Birds, *Trans. Carlisle Nat. Hist. Soc.* 9: 62.

BOLAM, G. C. (1912). *Birds of Northumberland and the Eastern Borders*, Alnwick, Blair.

BOOTH, E. T. (1881–7). *Rough Notes on the Birds Observed During Twenty-five Years Shooting and Collecting in the British Isles*, London, Porter and Dulau.

BRYAN, B. (1922). Dotterel in Staffordshire, *Brit. B.* 16: 55.

BURTON, R. E. (1946). Dotterel in Northamptonshire, *Brit. B.* 39: 93.

CAPEK, V. (1886). Aus dem Riesengebirge, *Mitth. Orn. Ver. Wien* 10: 241–2.

CHANCE, E. P. (1922). *The Cuckoo's Secret*, London, Sidgwick and Jackson.

CHISLETT, R. (1948). *The Birds of Yorkshire*.

CLAY, T. (1964). 'Ectoparasite' in *New Dictionary of Birds*, London.

COLLETT, R. and OLSEN, A. (1921). *Norges Fugle*, Oslo.

CONGREVE, W. M. and FREME, S. W. P. (1930). Seven Weeks in Eastern and Northern Iceland, *Ibis* for 1930: 193–228.

COOPER, J. (1861). Letter in *Zool.*, p. 7638.

CUMMINGS, S. G. (1917). Dotterel in North Wales, *Brit. B.*, 10: 189.

CURRY-LINDAHL, K. (1963). *Våra Faglar i Norden*, Stockholm, 2nd ed.

DARLING, F. and BOYD, J. M. (1969). *The Highlands and Islands*, London, Collins, 2nd ed.

DARWIN, C. (1874). *The Descent of Man*, 2 vols., London, 2nd ed.

DAVIS, P. E. and JONES, P. H. (1967). Welsh Report for 1967 (Dotterel, p. 8). *Nature in Wales*, 11: 1–16.

DEMENTIEV, G. P. *et al.* (1954). *Birds of the Soviet Union*, Moscow. Vol. 3 in Israel Program for Scientific Translations, Jerusalem, 1969 (Reference to Russian workers in section on Dotterel, pp. 57–63).

DHARMAKUMARSINGHI, K. S. (1945). The Bustard Quail at Home. (*Turnix suscitator taigoori*,) *Avic. Mag.* 10: 58–60).

FRANKE, H. (1952). Unser Mornellregenpfeifer, *Vogelkunde Nachr. aus Österreich* 1: 12–13. (1953). Zur Biologie des Mornellregenpfeifers, *Photographie und Forschung* 5: 200–6.

FORREST, H. E. (1927). Dotterel in Shropshire, *Brit. B.* 21: 46.

GARNETT, R. M. (1951). Dotterel in Norfolk in March, *Brit. B.* 24: 372.

GILROY, N. (1923). Observations on the Nesting of the Dotterel, *Ool. Rec.* 3: 1–7.

GORDON, S. P. (1915). *Hill Birds of Scotland*, London, Arnold. (1936). *Thirty Years of Nature Photography*, London, Cassell, pp. 65–8.

GUDMUNDSSON, F. (1951). The Effects of Recent Climatic Changes on the Bird Life of Iceland, *Proc.* 10 *Int. Orn. Cong.* 502–14.

GUICHARD, K. M. (1956). Observations on Wintering Birds near Tripoli, Libya, *Ibis* 98: 311–16.

GURNEY, J. H. (1921). Dotterel in Norfolk in May, *Brit. B.* 14: 257.

HAARTMAN, L. von., *et al* (1966). Pohjolan linnut värikuvin. I. - Otava, Helsinki.

HANF, B. (1889) in *Mitth. Naturw. Ver. Steierm* for 1889: 113.

HARRIS, M. P. (1967). The Biology of Oystercatchers . . . , *Ibis* 109: 80–93.

HARVIE-BROWN, J. A. and BUCKLEY, T. E. (1887). *A Vertebrate Fauna of Sutherland, Caithness and West Cromarty*, Edinburgh, Douglas. (1895). *A Vertebrate Fauna of the Moray Basin*, Edinburgh, Douglas, 2 vols.

HARVIE-BROWN, J. A. (1906). *A Vertebrate Fauna of the Tay Basin and Strathmore* Edinburgh, Douglas.

HARVIE-BROWN, J. A. and MACPHERSON, H. A. (1904). *A Vertebrate Fauna of the North West Highlands and Skye*, Edinburgh, Douglas.

HAVILAND, M. (1917). Notes on the Breeding Habits of the Dotterel on the Yenesei, *Brit. B.* 11: 6–11.

HEWITSON, W. C. (1846). *Coloured Illustrations of the Eggs of British Birds*, London, 2nd ed.

HEYDER, R. (1960). Die Südareale des Mornellregenpfeifers *Eudromias morinellus* (L.) in Europa, *Abn. u. Ber. Mus. Tierk.*, Dresden 25: 47–70. (1962). Nachlese zur Verbreitung und Biologie des Mornellregenpfeifers . . . , *ibid.* 26: 100–11.

HEYSHAM, T. C. (1830). *Magazine of Natural Hist.* 3. (1883). *Magazine of Natural Hist.*

HILDÉN, O. (1965). Zur Brutbiologie des Temminckstrandlaufers *Calidris temminckii, Orn. Fenn.* 42: 1–5. (1966). Über die Brutbeteiligung des Geschlechter beim Mornellregenpfeifer, *Orn. Fenn.* 4: 16–19. (1967). Lapin Pesimälinnusto Tutkimuskohteena, *Luonnon Tutkisa* 71: 152–62.

HOFFMANN, A. (1949). Über die Brutpflege des Polyandrischen wasserfasans *Hydrophasianus chirurgus, Zool. Jahrb. Abt. Syst. Ok. Geog. Tier.* 76: 367–403.

HÖHN, E. O. (1967). Breeding Biology of Wilson's Phalarope (*Steganopus tricolor*) in Central Alberta, *Auk* 84: 220–44. (1968). Some Observations on the Breeding of Northern Phalaropes, *Auk* 85: 316–27.

HOLDER, F. W. (1945). Dotterel in Lancashire, *Brit. B.* 38: 257.

HOLMES, R. T. (1966). Breeding Ecology and Annual Cycle Adaptations of the Red-backed Sandpiper *Calidris arctica* in Arctic Alaska, *Condor* 68: 3–46. (1970). Differences in Population Density, Territoriality and Food Supply of Dunlin on Arctic and Subarctic Tundra, *Animal Populations in Relation to their Food Resources* (ed. A. Watson), London and Edinburgh, Blackwell.

HOMES, R. C. (1947). Dotterels near Dungeness, *Brit. B.* 41: 125.

HUMPHREYS, G. R. (1919). Dotterel in Co. Dublin, *Brit. B.* 13: 61.

JENKINS, D. (1957). The Breeding of the Red-legged Partridge, *Bird Study* 4: 97–100.

JOHNS, J. E. (1964). Testosterone induced Nuptial Feathers in Phalaropes, *Condor* 66: 449–54.

JOHNS, J. E. and PFEIFFER, E. W. (1963). Testosterone induced Incubation Patches in Phalarope Birds, *Science* 140: 1225–6.

JOURDAIN, F. C. R. (1929). Section on Dotterel *in* Bent, A. C. (1929). (1940). Sections on 'Breeding', 'Food', and 'Distribution Abroad' *in* Witherby, H. F. (1938–41).

KENNEDY, P. J., *et al.* (1954). *The Birds of Ireland*, Edinburgh and London, Oliver & Boyd.

KIMAKOVICZ, M. von (1896). Zur Vögelfauna Siebenburgens, *Verh. u. Mitth. Siebenb. Ver. Naturw.* 45: 35.

KIRKMAN, F. B. (1937). *Bird Behaviour*, London, Nelson and Jack.

KRAMPITZ, H. (1940). Vom Mornellregenpfeifer (*Charadrius morinellus*) (L.), *Ber. Ver. Schles. Orn.* 25: 71–2.

LACK, D. (1966). *Population Studies of Birds*, Oxford, Clarendon Press. (1968). *Ecological Adaptations for Breeding in Birds*, London, Methuen.

LATHAM, J. (1824). *A General Synopsis of Birds*, London (Dotterel 3: 334–7).

LAVEN, H. (1949). Beitrage zur Biologie des Sandregenpfeifers *Charadrius hiaticula* (L.), *J. Orn.* 88: 183–287.

LEVIN, N. O. (1956). Life History Studies on *Porrocaecum ensicaudatum, Dissertation Abstracts* 16: 1969.

LAMB, H. M. (1966). *The Changing Climate*, London, Methuen.

LAMBERT, H. (1957). Birds of Greece, *Ibis* 99: 43–68.

LANCASTER, D. A. (1964). Life History of the Boucard Tinamou in British Honduras, *Condor* 66: 165–81, 253–76. (1964a). Biology of the Brushland Tinamou, *Bull. Amer. Nat. Hist.* 127: 269–314.

LARSON, S. (1957). The Suborder Charadrii in Arctic and Boreal Areas during the Tertiary and Pleistocene, *Acta Vertebratica* 1: 1–84.

LINTIA, D. (1955). *Păsările din Romîniei (R.P.R.)* 3: 253.

LONG, S. H. (1937). Dotterel on Holy Island, *Brit. B.* 31: 60.

LORENZ, K. (1966). *On Aggression*, London, Methuen.

LOWE, P. R. (1915). On Down Pattern of *Oreophilus ruficollis* chick, in *Ibis* for 1915: 339.

MACGILLIVRAY, W. (1852). *A History of British Birds*, London, 5 vols. (1855). *The Natural History of Deeside and Braemar*, London.

MACPHERSON, H. A. (1892). *A Vertebrate Fauna of Lakeland*, Edinburgh, Douglas.

MACPHERSON, H. A. and DUCKWORTH, W. (1886). *The Birds of Cumberland*, Carlisle.

MCVEAN, D. N. and RATCLIFFE, D. A. (1962). *Plant Communities of the Scottish Highlands*, London, H.M.S.O.

MARAN, J. (1946). Kulik hnědy *Charadrius morinellus* L.v. krkonosich, *Sylvia* 8: 49–53.

MEARES, C. S. (1917). Field Notes on the Nesting of Dotterel in Scotland, *Brit. B.* 11: 12–14.

MEINERTZHAGEN, R. (1921). Some Preliminary Remarks on the Velocity of Migratory Flight among Birds, *Ibis* 3: 228–38. (1954). *The Birds of Arabia*, Edinburgh, Oliver & Boyd (p. 484).

MERIKALLIO, E. (1958). Finnish Birds, Their Distribution and Numbers, *Fauna Fennica* 5: 1–181.

MERRIN, P. E. (1948). Dotterel in Derbyshire, *Brit. B.* 41: 156.

MILLER, A. H. (1931). Observations on the Incubation and the Care of the Young in the Jacana, *Condor* 3: 32–3.

MOLTENO, D. J. (1936). (Dotterel seen in Glen Lyon), *Scot. Nat.* 221: 134.

MOSGOVOI, A. and BISHAEVA, L. (1959). The Life-cycles of *Porrocaecum ensicaudatum* . . . , *Helminthologia Bratislava* 1: 195–7 (English and German summaries, p. 157).

NETHERSOLE-THOMPSON, C. and D. (1938–41). Contributions *in* Witherby, H. F. *et al.* (1942). Eggshell Disposal by Birds, *Brit. B.* 35: 162–9; 191–200; 214–23; 241–50. (1943). Nest-site Selection by Birds, *Brit. B.* 37: 70–4; 88–94; 108–13. (1961). 'The Breeding Behaviour of the British Golden Plover' and 'Breeding Behaviour of the Dotterel' *in* Bannerman, D. A., 10: 206–14; 246–53.

NETHERSOLE-THOMPSON, D. (1934). Some Aspects of the Territory Theory, *Ool. Rec.* 14: 15–23; 79–93. Contribution on Dotterel *in* Witherby H. F. *et al.* 4: 384–6; 5: 282. (1951). *The Greenshank*, London, Collins. (1957). Ecological Note on Golden Plovers in the Cairngorms, *Scot. Nat.* 69: 119–20. (1957a). 'Northern' Golden

Plovers in Northern Parts of Scotland, *Scot. Nat.* 69: 121–2. (1961). Notes on Nesting of Temminck's Stint in Spey Valley *in* Bannerman, D. A. 9: 287–8. (1961). 'The Breeding Behaviour and Breeding Biology of the Oystercatcher' *in* Bannerman, D. A. 10: 65–74 and 310–18. (1966). *The Snow Bunting*, Oliver & Boyd. (1969). Notes on Breeding Biology of the Dotterel *in* Darling, F. and Boyd, J. M. (2nd ed.), pp. 168–9. (1971.) *Highland Birds* (Inverness, H.I.D.B.).

NEW DICTIONARY OF BIRDS (1964), London and New York, Nelson (ed. A. L. Thomson).

NICOLL, C. S. *et al.* (1967). Prolactin and Nesting Behaviour in Phalaropes, *Gen. and Comp. Endocrin.* 8: 61–5.

NIETHAMMER, G. (1942). *Handbuch der Deutschen Vögelkunde* 3: 143–7, Leipzig.

OAKES, C. (1953). *The Birds of Lancashire*.

OLDHAM, C. (1932). Colour of Feet and Bill of Dotterel Chick, *Brit. B.* 25: 364.

ORING, L. W. (1969). Egg-laying of a Golden Plover, *Ibis*, 109: 434.

PARMELEE, D. F. *et al.* (1967). The Birds of South-eastern Victoria Island, *Bull.* 222 *Nat. Mus. Canada*, Ottawa.

PARMELEE, D. F. and MACDONALD, S. D. (1960). The Birds of West Central Ellesmere Island, *Bull. 169 Nat. Mus. Canada*, Ottawa.

PEARSON, A. K. and O. P. (1955). Natural History and Breeding Behaviour of the Tinamou *Northoprocta ornata*, *Auk*, 72: 113–27.

PENNANT, T. (1790). *A Tour in Scotland and Voyage to the Hebrides* (5th ed.), Chester. (1812). *British Zoology* (*Class 2: Birds*), London, White.

PERRY, R. (1948). *In the High Grampians*, London, Lindsay Drummond.

PEUS, F. (1957). Zur Kenntnis der Brutvögel Griechenlands, *Mitt. Zool. Mus. Berlin* 33: 227–78.

PITELKA, F. (1959). Numbers, Breeding Schedules, and Territoriality in Pectoral Sandpipers of northern Alaska, *Condor* 61: 233–64.

PLESKE, T. (1928). Birds of the Eurasian Tundra. *Memoirs of Boston Soc. Nat. Hist.*

PULLIAINEN, E. (1970). On the Breeding Biology of the Dotterel, *Orn. Fenn.* 47: 69–73 (1971). Breeding Behaviour of the Dotterel, *Charadrius morinellus*, Report 24, *Värriö Subarctic Research Station* (*Univ. Helsinki*).

RAINES, R. J. (1946). Dotterels in Notts., *Brit. B.* 39: 148. (1962). Birds in N.E. Greece in Summer, *Ibis* 104: 196.

RANDALL, T. E. (1959). (On Wilson's phalarope) *in* Bannerman, D. A. 9: 204–6.

RATCLIFFE, D. A. (1957). Dotterels nesting in Ross, *Scot. Nat.* for 1957: 124–5. (1962). Breeding Density in the Peregrine *Falco peregrinus* and the Raven *Corvus corax*, *Ibis* 104: 13–49. (1967). Conservation and the Collector in *The Biotic Effects of Public Pressures on the Environment*, the Third Scientific Staff Symposium held at Monks Wood Experimental Station, 20th–21st March, 1967.

RAYFIELD, P. A. (1948). (Dotterels in W. Ross), *Scot. Nat.* 60: 136.

RIDPATH, M. G. (1964). The Tasmanian Native Hen, *Aust. Nat. His.* 195: 14: 346–50.

RINGROSE, B. J. (1934). (Dotterels in Hampshire), *Brit. B.* 28: 175.

RITTINGHAUS, H. (1962). Untersuchungen zur Biologie des Mornell-regenpfeifers (*Eudromias morinellus*) in Schwedisch Lappland, *Zeitschrift für Tierpsychologie* 19: 538–58.

SANTER, E. (1922). Vögelkundliche Beobachtungen im Kärtner Nockgebiet, *Carinthia* 2111: 41–2.

SCHLOTT, M. (1938). Neuer Brutnachweis des Mornellregenpfeifers (*Charadrius morinellus* L.) aus dem Riesengebirge, *Ber. Ver. Schles. Orn.* 23: 31–2.

SCHAANNING, H. T. L. (1907). *Ostfinmarkens Fuglefauna.* Bergens Mus. Aarb.

SCHOLES, W. R. and CLARKE, S. H. (1951). Dotterels in Lancashire, *Brit. B.* 44–288.

SEATH, R. (1935). (Dotterels nesting in Grampians), *Scot. Nat.* 211: 148.

SEEBOHM, H. (1884). *A History of British Birds,* London, Porter 3: 30–4.

SELBY, P. J. (1833). *Illustrations of British Ornithology,* Edinburgh.

SETH-SMITH, D. (1905). (On Reversed Courtship) in *Proc. 4th Int. Orn. Cong. ser.* 5: 63–7.

SCHÖNWETTER, M. (1960–6). *Handbuch der Öologie,* ed. W. Meise, Berlin.

SIM, G. (1903). *A Vertebrate Fauna of Dee,* Aberdeen.

SIMMONS, K. L. (1953). Some Aspects of the Aggressive Behaviour of three closely-related Plovers, *Ibis* 95: 115–27. (1956). Territory in the Little Ringed Plover (*Charadrius dubius*), *Ibis* 98: 390–9. (1961). Further Observations on Foot Movements in Plovers and other Birds, *Brit. B.* 3: 54–419.

SIMSON, C. (1966). *A Bird Overhead,* London, Witherby.

SLUITERS, J. E. (1938). Bijdrage tot de Biologie van den Kleinen Plevier, *Ardea* 27: 123–51. (1948). Notes on the Breeding of the Little Ringed Plover, *Limosa* 21: 83–5.

SMITH, K. D. (1965). *The Birds of Morocco, Ibis* 107: 493–526.

SMITH, K. D. and ASH, J. S. (1958). Dotterels at Portland Bill, *Brit. B.* 41: 28.

SOLLIE, J. F. (1961). Twee Broedgevallen van de Morinelplevier (*Charadrius morinellus*) in de Noordoostpolder, *Limosa* 34: 274–6.

SOULSBY, E. J. L. (1962). 'Endoparasite' in *New Dictionary of Birds.*

STEINIGER, F. (1959). *Die Grossen Regenpfeifer,* N.-Brehm Buch. No. 240.

STEIN-SPIESS, S. (1959). Mornellregenpfeifer im Zibinsgebirge, *Aquila,* 66: 275–309.

STERBETZ, I. (1957). Dotterels nesting in the southern Carpathians, *Aquila,* 63/64: 338. (1959). A Havasi Lile (*Charadrius morinellus* L.) Madyaronzágon, *Allatani Közlemények* 47: 143–7.

SWAN, M. (1950). Dotterel possibly nesting in W. Ross, *Scot. Nat.* 62: 184.

THORNTON, T. (1804). *A Sporting Tour Through the Northern Parts of England and Great Part of the Highlands of Scotland*, London, Sporting Library.

TICEHURST, N. F. (1929). *A History of the Birds of Kent*, London, Witherby (Dotterel, pp. 415–17).

TINBERGEN, N. (1935). Field Observations of East Greenland Birds, I. The Behaviour of the Red-necked Phalarope in Spring, *Ardea* 24: 1–42. (1953). *The Herring Gull's World*, Collins. (1959). The Behaviour of the Red-necked Phalarope on its Breeding Grounds, *in* Bannerman, D. A. 9: 198–9.

USPENSKII, S. M. *et al.* (1962). Birds of North-east Yakutia, *Ornitologya* 4: 64–86 (in Russian).

VAN ELBURG, H. and VAN den BERG, A. (1967). Het Voorkomen van den Morinelplevier (*Charadrius morinellus*) in de IJsselmeerpolders in de jaren 1961 t/m 1966 *Intern Rapport* 72, Rijksdienst voor de IJssel-meerpolders, Zwolle. (1969) *ibid*, and later tables.

VAUGHAN, R. (1952). Accertata Nidificazione sul Massiccio della Maiella (Abruzzi) del Piviere tortolino *Charadrius morinellus* (L.), *Rivista Ital. Orn.* 22: Brevi nota. (1953). Alcune Osservazioni su gli Uccelli del Massiccio della Maiella (Abruzzi), *Rivista Ital. Orn.* 23: 137–42.

VENABLES, L. S. V. and VENABLES, U. M. (1955). *Birds and Mammals of Shetland*, Edinburgh, Oliver & Boyd.

VINCE, M. A. (1964). Synchronization of Hatching in American Bobwhite Quail, *Nature* 203: 1192–3. (1966). Potential Stimulation Produced by Avian Embryos, *Anim. Behav.* 14: 34–40.

VOOUS, K. H. (1960). *Atlas of European Birds*, London, Nelson.

WALKINSHAW, L. H. (1903). Some Life History Studies of the Stanley Crane, *Proc. Int. Orn. Cong.* 13: 344–53.

WALPOLE-BOND, J. (1932). Colour of Feet and Bill of Dotterel Chick, *Brit. B.* 25: 337. (1938). *A History of Sussex Birds*, London, Witherby, 3 vols. (Dotterel 3: 122–4).

WATSON, A. (1955). Dotterels in Wester Ross, *Scot. Nat.* 67: 113. (1965). A Population Study of Ptarmigan (*Lagopus mutus*) in Scotland, *J. Anim. Ecol.* 34: 135–72. (1966). Hill Birds of the Cairngorms, *Scot. Nat.* 4: 178–203 (Dotterel, pp. 184–6).

WATSON, A. and JENKINS, D. (1968). Experiments in Population Control by Territorial Behaviour in Red Grouse, *J. Anim. Ecol.* 37: 595–614.

WATSON, A. and MOSS, R. (1970). 'Population Limitation in Vertebrates' in *Animal Populations in Relation to their Food Resources*, Oxford and Edinburgh, Blackwell.

WATSON, J. (1888). The Ornithology of Skiddaw, Scafell and Helvellyn, *Naturalist*. (1888–9). The Northern Distribution of the Dotterel, *Westmorland Nat. Hist. Record* 6: 162–4 and 176–9.

WEIR, T. (1960). The Fool of the Moor, *Scot. Field*, May, 1960: 45–8.

WHISTLER, H. (1941). *Popular Handbook of Indian Birds*, 3rd ed., Edinburgh, Oliver & Boyd.

WILSON, J. O. (1933). *The Birds of Westmorland and Northern Pennines* London, Hutchinson.

WITHERBY, H. F. *et al. The Handbook of British Birds*, 5 vols., London, Witherby. (Dotterel 4: 384–8 and 5: 282–303.)

WOLLEY, J. (1905). *Ootheca Wolleyana*, London, Porter.

WYKES, J. C. (1937). Dotterels near Edinburgh in May, *Scot. Nat.* 227: 144.

WYNNE-EDWARDS, V. C. (1962). *Animal Dispersion in Relation to Social Behaviour*, Edinburgh, Oliver & Boyd.

YARRELL, W. (1841). *A History of British Birds*, London, Van Voorst. (1882–4). London. (4th ed. revised and enlarged by A. Newton.)

YEATES, G. K. (1947). *Bird Haunts in Northern Britain*, London, Faber and Faber.

Acknowledgments

You cannot write a monograph without a lot of help. Until 1950 Carrie shared my camps and field work. Afterwards Brock worked with me.

In 1966 the Natural Environment Research Council awarded me a research grant. I warmly thank Professor V. C. Wynne-Edwards, F.R.S., my supervisor, for his encouragement and stimulating advice. I also learnt much from discussions at the Unit of Mountain and Moorland Ecology and Culterty Field Station.

The Dotterel owes much to Adam Watson, who has been my constant, kindly critic and adviser. He has contributed many unpublished notes and observations, read chapters 4–14 in draft, and, with Ray Parr, made a statistical analysis of my data. I also thank Bob Moss for his statistical interpretation of our clutch-size records.

From Colin Murdoch's freehand sketches, embodying our joint data for 1953–4, Nick Picozzi drew the diagrams showing maximum breeding density.

Dr Derek Ratcliffe contributed many observations and wrote the detailed chapters on the dotterel's breeding habitat in Britain and its status in England. He is also responsible for descriptions of vegetation at nests and read chapters 4–14 in draft. I gratefully acknowledge all this help.

The late Ernest Blezard greatly helped me with unpublished notes, records and observations. With Derek Ratcliffe he is largely responsible for Appendix 4.

Ian Pennie kindly allows me to use his fine colour photographs as a frontispiece and has shared the cost of the block with me.

I am also grateful to Miss M. Garnett, who kindly gave me extracts from the diaries of Francis Nicholson and J. F. Peters.

For the use of other unpublished data I particularly thank H. Auger, Dr H. M. S. Blair, S. E. Cook, Dr G. Franklin, Seton Gordon, C.B.E., Dr O. Hildén, Prof. W. Hobson, D. Humphrey, J. Markham, W. Marshall, C. Murdoch, Prof. P. Palmgren, R. Perry, J. F. Sollie, T. E. Randall, A. Tewnion, G. L. Trafford, S. Van den Bos, A. Watson, Sen., D. Watson and G. K. Yeates.

Many friends at home and abroad have greatly helped me.

Britain: Those already mentioned and W. Austin, David Bannerman, O.B.E., S. Batchelor, C. Best, E. Bowser, Prof. Rogers Brambell, R. H. Brown, J. Burton, Dr B. Campbell, S. S. Chesser, late R. Chislett, John Christian, R. Collier, R. Cranna, Dr Jack Dainty, Dr F. H. Day, G. Douglas, M.C., Prof. G. Dunnet, M. Everett, late T. Gordon, M. J. P. Gregory, M. Haas, late Canon Aidan Hervey, R. Hewson, I. E. and M. Hills, G. Horne, D. Hulme, R. Laidler, H. H. Lamb, A. Lance,

F. Mackintosh, H. MacPherson, J. MacRae, late P. M. Meeson, late Lt.-Col. R. Meinertzhagen, D.S.O., D. Merrie, P. Merrin, Dr H. Milne-Redhead, Dr N. Moore, R. Moreau, late Prof. W. H. Pearsall, Dr I. Pennie, Prof. D. Poore, J. Robson, H. Shurrock, A. Smith, I. Smith, R. Stokoe, Margaret Suggate, Doreen Tewnion, W. Thompson, J. Tomkinson, N. Usher, R. Wagstaffe and Tom Weir.

U.S.S.R. Professor A. I. Ivanov, Curator of Birds, Zoological Institute, Leningrad, kindly wrote the detailed accounts of the dotterel's distribution and habitat in the Soviet Union.

Finland: Professor P. Palmgren, Dr O. Hildèn, T. Stjernberg, and Dr E. Pulliainen.

Sweden: U. Houmann and P. O. Swanberg.

Norway: Konservator E. Barth and Drs Y. Hagen and H. Holgersen. Dr Hugh Blair is largely responsible for accounts of distribution and habitat in Scandinavia.

Denmark: Professor Hans Johansen and Dr F. Salomonsen.

Holland: N. Ruitenberg, J. F. Sollie and Professor K. Voous.

Germany: Professor E. Stresemann, R. Heyder, H. Rittinghaus and N. Ruitenberg.

Austria: H. Franke.

Czechoslovakia: F. J. Turček.

Italy: R. Vaughan.

North America: Professors Dave Parmelee and Dick Phillips, and Walter Graul, University of Minnesota. Dr E. O. Höhn, University of Alberta and Dr R. T. Holmes, Dartmouth College, New Hampshire, and R. B. Weeden.

INSTITUTIONS

British Museum (Natural History): The late Sir Norman Kinnear and J. D. MacDonald, former, and Dr David Snow, present, Heads of the Bird Room. I also thank Derek Goodwin and Colin Harrison and acknowledge the data-files of the E. P. Chance and other collections.

Royal Scottish Museum, Edinburgh: Rodger Waterston, Keeper of Zoology, L. Lyster, Dr A. S. Clarke and the late James Dunbar.

Edward Grey Institute of Field Ornithology: David Lack, Director, for many kindnesses and Miss J. Coldrey for copying extracts from Arthur Whitaker's Diaries.

British Trust for Ornithology: H. Mayer-Gross and the Research Committee for data from the Nest Records Scheme.

Booth Museum, Brighton: J. Morley, M.A., Director, for allowing me to make extracts from E. T. Booth's MS diaries. R. G. Hiles, in charge of this unique museum, greatly helped me.

Libraries: Marischal College, Aberdeen; Edward Grey Institute, Oxford; King's College, Edinburgh; and Inverness, Ross and Sutherland County Libraries. Donald Anderson, Librarian, Inverness County Library, has been a tower of strength.

Translations: Dr Lil de Kok translated sections of H. Rittinghaus's important paper. Ronald MacDonald, Director of Education, Inverness, arranged for a translation of H. Franke's papers. I warmly acknowledge his help and that of Dr D. J. MacDonald, formerly Rector of Inverness Royal Academy. Teunis van Gelderen kindly translated an important Dutch paper for me.

Maps: Nick Picozzi kindly drew the maps of the dotterel's breeding distribution. For the map of distribution abroad he used Voous's *Atlas of European Birds*, with additions to bring it up to date and for winter distribution G. P. Dementiev's *Birds of the Soviet Union.*

For the other maps Adam Watson plotted all the locations from my collected data, with assistance from David Welch for some English breeding sites. From these roughs Nick Picozzi produced the final maps.

I thank Professor K. H. Voous for allowing me to use his map and data.

Proofs: Doctors Ratcliffe and Watson read their own contributions in proof and Dr Hugh Blair read other chapters.

I thank those on whose land I have worked, particularly Forestry Commission (Northern Conservancy) and David Robertson, formerly Head Forester, Queen's Forest, Glenmore, the Nature Conservancy, Lt.-Col. Ian Grant, M.B.E., late Major Tom Drake, C. MacFarlane-Barrow and Mrs Marjorie Fergusson.

I remember with gratitude my mother for her constant faith in my work and Francis Jourdain, Frank Kirkman, Julian Huxley, Cecil Stoney, John Walpole-Bond and Harry Witherby for their inspiration.

Finally, without Maimie's encouragement, I would never have finished this book.

List of Tables

TABLE I   Time schedule (in days) of some Boreal and Arctic Waders

Species	Egg-laying	Incubation	Fledging	Total	Egg-size (mm)
Dotterel	3 - 5	23.5 - 30	25 - 30	52 - 65	41.1 x 28.8 (100)
American Golden Plover	c.7	26	22	c.55	48.25 x 33.12 (24)
Ruddy Turnstone	4 - 5	21 - 22	c.19	44 - 46	41.17 x 28.65 (22)
Purple Sandpiper	4 - 5	22 - 23	c.21	45 - 49	37.3 x 26.2 (100)
Baird's Sandpiper	4 - 5	21 - 22	c.20	45 - 47	34.18 x 24.3 (60)
White-rumped Sandpiper	c.4 - 5	22	c.15 - 17	41 - 44	33.26 x 29.53 (54)
Semi-palmated Sandpiper	c.4 - 5	18 - 19	14	37 - 38	30.76 x 21.82 (35)
Sanderling	c.4 - 5	24.2 - 31.6	17	up to 54	35.32 x 24.61(100)
Grey Phalarope	c.4 - 5	18 - 19	16 - 20	38 - 44	32.17 x 22.3

TABLE 2   NEST DANCES

Species	Pattern	Sexual behaviour
Red-necked phalarope	At first ♀ makes scrapes independently. After copulation ♀ leaves water and starts scraping. ♂ follows and begins to make incomplete scrape-movements. Periodically uses same calls as made in 'ceremonial flights'. ♂ and ♀ next make scrapes independently, but after a few days join in 'sideways building' at same scoop and carry out 'symbolic' nest-relief behaviour. ♀ finally chooses scoop in which she lays her first egg. (N. Tinbergen, 1959).	Copulation not recorded during nest-dances.
Wilson's phalarope	♀ apparently forms several scrapes, one of which she finally selects as nest. No nest-dances recorded. (E.O. Höhn).	Copulation not recorded in context of scrape-making.
Oystercatcher	Mutual nest-dances. Established pairs sometimes 'scamp' these displays and use same scoop in several consecutive years. (D.N.-T., 1961).	Copulation not a normal component of nest-dances, but ♂ seen to tread ♀ sitting on one egg.
Ringed plover	♂ regularly shows ♀ his chosen scrape, tilting forward on breast with wings pointing upwards. ♂ also sometimes broods on empty scrape. (J.K. Stanford).	Nest-dances sometimes lead to copulation. (D.N.-T.)
Little ringed plover	Sometimes arrive as pairs; but ♂ forms or enlarges a hollow, leaning forward and rotating on ground, body and tail slanting up, feet scratching backwards. Simultaneously ♂ opens his tail to show off brown, black and white colour pattern. The conspicuous movements of his tail probably attract a hen. When ♀ approaches scraping ♂ he steps out of hollow, stands on rim with his back to ♀ and fans his tail over scrape. ♀ now steps	Nest-dances sometimes die down or end in copulation.

continued

D.   TABLE 2   *continued*

Species	Pattern	Sexual behaviour
Little ringed plover *(continued)*	under his tail and into scoop. ♂ meanwhile moves slowly away, tail fully spread, simultaneously jerking little stones towards nest. Tail relaxes as he gets further away from nest. ♀ meanwhile stays in and rotates in nest or moves away. This pattern is sometimes repeated. (K.L. Simmons).	
Lapwing	After incipient scrape movements in the flock, ♂ takes territory, making many scrapes fully a fortnight before ♀ approaches him. At first his 'rocking displays' are incomplete and unco-ordinated, but finally he establishes firm display-centres from which he periodically makes song-flights. As ♀ slowly approaches, ♂, rocking on scrape, jerks straws over shoulders. Then, tilting forward, head down and tail up, he slowly moves away from ♀, jerking bits and pieces over shoulders. If ♀ leaves, display-chain is broken. Later ♀ joins in nest-dances and ♂ upends beside her. (D.N.-T., 1961).	During nest-dances ♂ frequently mounts ♀ but these do not always lead to firm pairs.
Sanderling	♀ settled in scrape with head low, tail almost vertical' and 'squirming' a little. ♂ then ran in, thrust his bill down beside her, and removed bits of lining from scrape. ♂ then stood on ♀'s back, stepped off, and deliberately eased her out of the scoop with his bill and forehead thrust beneath her belly. Both birds then ran off side by side, bodies pressed together and rubbing vigorously. About 10 feet further on ♂ stopped ♀ by pressing his bill against her breast. ♂ then mated ♀. (D.F. Parmelee, 1970).	Pairs apparently do not lay eggs in these pre-copulatory scrapes.

Q

TABLE 3 NESTS AND NEST-SITES (D.A. Ratcliffe)

District	Situation	Nest
1. Central Grampians	Large boulder 3 yards away, but no stones near nest which was in middle of 'polygon hummock' in continuous deep carpet of *Rhacomitrium lanuginosum* thrown into hummocks, with a few dwarf blaeberry *Vaccinium myrtillus* shoots round its edge. *Carex bigelowii* and *Vaccinium vitis-idaea* were only other plants growing nearby.	Mere depression in fringe-moss mat, lined dead blaeberry leaves and lichen *Cladonia uncialis*.
2. Central Grampians	Nest against flat sideways-tilted stone 3-4 inches high. Vegetation mixed: *Rhacomitrium lanuginosum* and *Deschampsia flexuosa*. (Not pure moss carpet).	Lined with *Cladonia uncialis* and *Cladonia sylvatica*.
3. Central Grampians	Very eroded *Festuca ovina* – *Rhacomitrium lanuginosum* ground.	Lining of *Cetraria islandica* and *Cladonia sp.*
4. E. Grampians	Vegetation mixed, but mainly lichen heath dominated by *Cladonia sylvatica* and *C. rangiferina* with a little *Rhacomitrium lanuginosum* and a good deal of *Carex bigelowii*, *Vaccinium myrtillus* and *Dicranum fuscescens*. Much of adjacent ground was *R. lanuginosum* heath, but nest not in this.	Lining a pad of *Cladonia sylvatica* with some *C. uncialis*.
5. E. Grampians	Nest lying right against 2 ft. length of old fence post and only a few feet from fence itself. Vegetation: *Carex bigelowii*, *Festuca ovina*, *Dicranum fuscescens*, *Cetraria islandica*, *C. sylvatica*, *C. rangiferina* (lichens rather sparse).	Lining of *Carex bigelowii* leaves with *C. sylvatica* *C. rangiferina* and *C. uncialis*.
6. E. Grampians	Mixture of *Rhacomitrium-Carex bigelowii* heath, with some *Nardus* and *Dicranum fuscescens* – *Carex bigelowii*.	Chicks about 2 days old.
7. E. Grampians	Only 10 yards from old wire fence. 3-4 bedded stones within 2 feet. Near edge of stony terrace above some *Nardus* beds. Birds had ignored extensive hummocky *Rhacomitrium* heath lower down spur.	Scrape in *Carex bigelowii*-*Dicranum fuscescens* with some *Gymnomitrium* crust. and lichens.

continued

TABLE 3 continued

District	Situation	Nest
8. E. Grampians	Low mound of hummocky mixed heath composed of *Carex bigelowii*, *Vaccinium myrtillus*, *V. vitis-idaea*, *Cladonia sylvatica*, and *C. rangiferina*.	Shallow scrape lined with sparse *C. sylvatica* and *C. uncialis*.
9. E. Grampians	In fairly dense *Carex bigelowii* with some *Rhacomitrium lanuginosum*.	Lining, a pad of *Cladonia uncialis* and some *R. lanuginosum*.
10. E. Grampians	Nest of low hummock on slight slope with short *Festuca vivipara*, *Lycopodium selago*, *Carex bigelowii*, sparse *Juncus trifidus*, *Galium hercinum*, *R. lanuginosum* and *C. uncialis*.	
11. E. Grampians	Nest beside bedded stone on hummocky ground, with mixed *Rhacomitrium – Carex bigelowii* community containing sparse *Nardus*. Scrape in *R. lanuginosum*, *C. bigelowii*, *D. flexuosa*, *Alchemilla alpina* and mixed lichens.	Lined with lichen pad mainly of *C. uncialis* and *C. squamosa*.
12. W. Ross	Nest scrape in stony *Rhacomitrium lanuginosum* carpet, protected on W. side of stone 5 inches high. Another stone lay on other side. *Rhacomitrium* carpet fairly continuous, with *Silene acaulis* and *Armeria maritima*.	Lined with dead leaves of *Salix herbacea* and *Cladonia uncialis*.
13. W. Ross	Nest on stony slope (c.10°). Ground of unstable eroded type affected by solifluction and soil with much bare stone. Open patchy vegetation with much *Juncus trifidus*. Nest in little patch of *Rhacomitrium lanuginosum* with large stone to one side.	Nest unlined; but *Carex bigelowii* and *Alchemilla alpina* grew on edge and small patches of *Silene acaulis* and *Armeria* within inches of it.
14. Lakeland, England	Scrape between two tilted stones amongst *Festuca ovina*, *Rhacomitrium lanuginosum* and dwarf *Vaccinium myrtillus*.	Lined with *Rhacomitrium*, *Cetraria islandica* and *Cladonia uncialis*.

continued

TABLE 3 *continued*

District	Situation	Nest
15. Lakeland, England	Nest beside 3 large stones near edge of densely block-strewn area in a community of *Rhacomitrium lanuginosum* and *Festuca ovina* with sparse *Vaccinium myrtillus*.	Lined *Rhacomitrium* only.
16. Lakeland, England	Placed between two bedded stones in very short dense *Festuca ovina* with a very little *Rhacomitrium lanuginosum* in small tufts and with *Polytrichum alpinum* as the only other plant.	Lining was a thick pad of *Cetraria islandica* which had certainly been carried in from outside and also a few scraps of *Polytrichum alpinum*, *Rhacomitrium lanuginosum* and *Festuca ovina*.
17. Lakeland, England	The nest was away from the very stony hummocks, and was on a low grassy mound with no exposed stones. It was in a close-grazed sward of *Festuca ovina*, with some *Rhacomitrium lanuginosum*, *Polytrichum alpinum*, *Cetraria islandica*, *C. aculeata*, *Cladonia uncialis* and *C. gracilis*.	Lining was a good pad of *Cladonia uncialis* with some *Cetraria islandica*. The altitude was 2900 feet.
18. Lakeland, England	Nest between stones lying flat in mixed *Festuca ovina* – *Rhacomitrium lanuginosum*.	Lining, dense pad of *Cladonia uncialis* with some *R. lanuginosum*.
19. Lakeland, England	In short *Festuca ovina* turf, between stones.	Lining of chopped *F. ovina*, *Cladonia uncialis*, *C. pyxidata* and *Cetraria islandica*.

continued

NESTS AND NESTSITES

District	Situation	Nest	Authority
Central Cairngorms	Nest deepish hollow in tussock of grass between two granite stones, embedded in stony ridge.	Nest lined grass and small pieces of lichen.	D.N.-T.
Central Cairngorms	Nest on granite block-strewn shoulder of rounded hill. Nest almost beside large granite rock. Eggs actually lying in hollow rock.	Sparingly lined with moss, lichens, small stones and a few mountain hare and ptarmigan droppings.	D.N.-T.
Central Cairngorms	Nest beside an embedded white quartz block not far from carpet of *Rhacomitrium* moss, but in grassy tump. Many large rocks and innumerable granite stones on ridge.	Lined moss.	D.N.-T.
Central Cairngorms	In a screefield of quite large granite rocks *c*.40 yards from snowfield.	Lined grass or moss, lichen round edge of cup.	D.N.-T.
Central Cairngorms	Between two smooth and rounded granite slabs on stony terrace.	No lining at all.	D.N.-T.
Central Cairngorms	Beside flat triangular rock on stony hillside.	Lined crowberry leaves.	D.N.-T.
Central Cairngorms	In grassy island among rounded blocks of granite.	Lined grass and moss. Two small feathers in cup.	D.N.-T.
Central Cairngorms	On plateau top in an area with much bare gravel, patches of moss and *Juncus trifidus* and *Salix*. In a hollow in a patch of *Juncus trifidus* turf alongside a granite stone which gave shelter on one side. The turf included *Salix herbacea*.	Lined with dead leaves of *Salix herbacea* (1967 leaves) and dead stems and leaves of *Juncus trifidus* with a few small bits of dead *Rhacomitrium lanuginosum* and *C. islandica*. These two species must have been taken to the nest. There were none growing within a foot of it.	A. Watson

continued

NESTS AND NEST-SITES *continued*

District	Situation	Nest	Authority
East Cairngorms	On gravelly ground with numerous boulders and open mosaic of vegetation. Nest between three stones in patch of *Rhacomitrium lanuginosum* with *J. trifidus*, *C. bigelowii* and *Festuca vivipara*.	Lining of *R. lanuginosum*, scraps of *C. bigelowii* leaf, *Cetraria islandica* and *Cladonia uncialis*.	D.A. Ratcliffe

NEST SITES

Notes made by D.A. Ratcliffe on vegetation shown
in a selection of photographs

A. Central Cairngorms, 1940

Nest in patch of mountain sedge, *Carex bigelowii* – wavy hair grass, *Deschampsia flexuosa* sward, with moss campion, *Silene acaulis* and Iceland moss, *Cetraria islandica*. Adjacent, open gravelly soils have much three-pointed rush, *Juncus trifidus*.

B. Central Cairngorms, 1965

Nest in patch of three-pointed rush, *Juncus trifidus* – wavy hair grass, *Deschampsia flexuosa* sward, with woolly fringe-moss, *Rhacomitrium lanuginosum* and *Cladonia sylvatica*.

C. West Cairngorms, 1945

Nest in carpet of woolly fringe-moss, *Rhacomitrium lanuginosum*, with mountain sedge, *Carex bigelowii* – wavy hair grass, *Deschampsia flexuosa* and *Cladonia uncialis*.

D. Central Cairngorms, 1958

Nest in carpet of woolly fringe-moss, *Rhacomitrium lanuginosum*, with mountain sedge, *Carex bigelowii* – wavy hair grass, *Deschampsia flexuosa*, *Cladonia sylvatica* and *C. uncialis*.

continued

E. Central Cairngorms, 1953

Nest in carpet of prostrate heather, *Calluna vulgaris* with *Cladonia sylvatica*.

F. West Grampians, 1968

Nest in sward of short tufted hair grass, *Deschampsia caespitosa* and mountain sedge, *Carex bigelowii*.

G. Sutherland, 1967

Nest in prostrate heather *Calluna vulgaris* mat with *Carex pilulifera* and wavy hair grass, *Deschampsia flexuosa*.

H. Drumochter, *c*.1900

(Title page of A Fauna of the Tay Basin and Strathmore, 1906, J.A. Harvie-Brown). Nest in woolly fringe-moss, *Rhacomitrium lanuginosum*, sheeps fescue, *Festuca ovina*, wavy hair grass, *Deschampsia flexuosa*, mountain sedge, *Carex bigelowii* and Blaeberry, *Vaccinium vitis-idaea*.

I. West Cairngorms

(Colour plate in Mountains and Moorlands, New Naturalist, 1950, W.H. Pearsall). Carpet of woolly fringe-moss *Rhacomitrium lanuginosum*, with sheeps fescue, *Festuca ovina* and mountain sedge, *Carex bigelowii*.

TABLE 4   EGG-LAYING BEHAVIOUR

Location	Date	Eggs	Behaviour	Authority
Cairngorms	13.6.34	2	1205 2 eggs in nest, 1250 - 1455 ♀ brooded eggs. Periodically preened herself and adjusted material on side of nest. 1455 ♂ changed places with ♀ and commenced brooding. Still 2 eggs.	C. & D.N.-T.
	14.6.34	3	1655 ♂ brooding.	
Cairngorms	20.6.49	2	1445 ♀ sitting on 2 eggs. ♂ not present. 2nd egg laid by 1705. ♀ sitting.	D.N.-T.
		3	2120 ♂ sitting.	
Cairngorms	2.6.39 3.6.39 4.6.39	1 1 2	♂ continuously brooding. ♂ sitting all day on first egg. ♂ sitting. 2nd egg now laid.	Seton Gordon
Grampians	5.6.23	2	1025 bird (sex?) sprang at 40 yards from nest with 2 eggs. Only 1 bird at nest.	N. Gilroy
		3	1300 bird (sex?) sitting on 3 eggs.	
Grampians	1.6.54	2	1130 ♂ brooding 2 eggs, ♀ nearby.	R. Perry
		3	1530 ♂. brooding 3 eggs.	
Grampians	25.5.67	1	"A rather brightly coloured bird" sitting on nest containing 1 egg. This bird picked up lichen scraps and threw them over its shoulder. 2nd scrape a few yards away. ♂ now copulated with ♀ a few yards from nest.	C. Murdoch
Nordoost Polder, Holland	5.5.61 8.5.61 10.5.61	1 2 3	No observations.	J.F. Sollie

continued

TABLE 4   *continued*

Location	Date	Eggs	Behaviour	Authority
Zirbitzkögel Austria	1.6.49 2.6.49 3.6.49	1 2 3	Pair close to nest. Pair close to nest. 0730, 3 eggs ♂ sitting, ♀ about.	H. Franke
Zirbitzkögel Austria	11.7.49 12.7.49 13.7.49	1 2 3	One bird brooding, mate standing beside it. 1745, ♀ changed with ♂ and laid 3rd egg. ♀ then left nest and disappeared.	H. Franke
Zirbitzkögel Austria	1.6.52 3.6.52	1 3	Extra ♂ appearing at nest where pair had nest with 1 egg. 0505 ♂ left 2 eggs on which he was brooding. 0515 ♀ took over. By 0630 3rd egg was laid. 0635 ♀ left nest.	H. Franke
Sweden	13.7.44 14.7.44	2	2nd egg laid between 1430 and 2030. 2045 - 2145 ♂ brooding. 0730 - 0855 ♂ brooding 2 eggs; ♀ always close to nest.	P.O. Swanberg
Abisko Sweden	2.7.59 3.7.59 3.7.59 4.7.59 5.7.59	1 1 2 2 3	1403 ♂ brooding, 1 egg cold. 1400 ♂ brooding, 1 egg warm. 1610 pair trapped and ringed near nest. 1745 2 eggs in nest. Both cold. Pair absent. 1215 ♂ sitting on 2 warm eggs. 1400 eggs warm, pair not seen. 1425 ♂ sitting on 3 eggs.	H. Rittinghaus
Värriötunturi Finland	6.6.69	1	At 1000 ♂, sitting on scrape, threw small pieces of plants backwards with bill and occasionally stretched head and neck backwards and forwards, and upwards. ♀ then came to nest and ♂ stood up. ♀, now very nervous, ran 30 yards from nest and ♂ again squatted. Finally ♀ returned and laid egg 1115.	E. Pulliainen

TABLE 5   INTERVALS BETWEEN EGGS (in hours)

	24-30	31-36	37-48	over 48	Total
Between 1st and 2nd egg	9	2	1	1	13
Between 2nd and 3rd egg	6	1	1	1	9

TIMES OF LAYING

1st egg (3)   c.1430, c.1500, c.1530

2nd egg (5)   1130 - 1530, "afternoon", 1430 - 2030, 1651 - 1942, c.1815

3rd egg (11)   0515 - 0630, c.0700, 0917 - 1053, 1050 - 1300, c.1130, c.1245, 1209 - 1355, 1130 - 1430,
                 1445 - 1705, c.1809, c.1930

TABLE 6   LAYING SEASON

Year	1st egg in earliest recorded clutch (Repeat clutches excluded)			1st egg in latest recorded clutch (Repeat clutches excluded)		
	Cairngorms	Central Grampians	Lakeland, England	Cairngorms	Central Grampians	Lakeland, England
1898	–	19 May	–	–	–	–
1908	–	16 May	19 May	–	4 June	6 June
1910	–	21 May	23 May	–	5 June	–
1911	–	14 May	11 May	–	4 June	19 May
1912	–	21 May	26 May	–	–	–
1914	20 May	23 May	29 May	–	–	–
1915	–	28 May	1 June	–	2 June	–
1917	–	–	17 May	–	–	–
1918	–	–	30 May	–	–	–
1920	12 May	–	15 May	–	–	25 May
1921	–	–	15 May	–	–	–
1922	25 May	–	24 May	28 May	–	–
1925	22 May	19 May	27 May	–	–	–
1926	27 May	27 May	26 May	–	–	–
1929	28 May	15 May	–	–	27 May	–
1931	–	22 May	–	–	–	–
1932	13 June	–	–	–	–	–
1933	12 May	5 June	–	5 June	–	–
1934	1 June	27 May	–	11 June	–	–
1935	after 5 June	–	–	–	–	–
1936	24 May	–	–	11 June	–	–
1937	20 May	–	–	27 May	–	–
1938	5 June	–	–	after 12 June	–	–
1939	31 May	–	–	5 June	–	–
1940	9 May	22 May	–	15 June	1 June	–
1941	27 May	–	–	7 June	–	–

continued

TABLE 6 *continued*

	1st egg in earliest recorded clutch (Repeat clutches excluded)			1st egg in latest recorded clutch (Repeat clutches excluded)		
Year	Cairngorms	Central Grampians	Lakeland, England	Cairngorms	Central Grampians	Lakeland, England
1942	28 May	19 May	–	3 June	–	–
1945	29 May	–	–	6 June	–	–
1946	30 May	–	–	–	–	–
1947	28 May	–	–	–	–	–
1948	10 June	–	–	18 June	–	–
1949	18 May	28 May	–	6 June	3 June	–
1950	27 May	25 May	–	after 30 June	5 June	–
1951	after 21 June	22 May	–	–	23 May	–
1952	–	22 May	–	–	–	–
1953	–	19 May	–	–	–	–
1954	6 June	22 May	–	21 June	7 June	–
1955	29 May	20 May	–	–	–	–
1956	30 May	–	–	after 17 June	–	–
1957	27 May	–	–	11 June	–	–
1958	31 May	23 May	–	–	–	–
1959	28 May	23 May	28 May	30 May	–	–
1960	22 May	22 May	25 May	8 June	–	–
1961	22 May	23 May	–	11 June	–	–
1962	3 June	26 May	–	–	–	–
1963	26 May	29 May	–	–	–	–
1964	23 May	16 May	–	–	–	–
1967	–	25 May	–	–	–	–
1968	–	–	25 May	–	–	–
1969	13 May	–	21 May	–	–	c.2 June
1970	–	–	14 May	–	–	c.16 May
Average date	27 May (36 years)	22 May (30 years)	23 May (18 years)	10 June (20 years)	3 June (10 years)	26 May (5 years)

TABLE 7 Laying Season of Dotterel and Ptarmigan in Cairngorms, Scotland

Early Years

Year	Mean temperature (°C) for May at Braemar (1113 ft)	Dotterel		Ptarmigan	
		1st egg	Comments	1st egg	Comments
1933	8.7	12 May	Very early season. Latest hen laid first egg 3 June.	6 May	An early year.
1936	8.4	24 May	Hens nested in waves. Last hen laid first egg on 11 June.	13 May	On same slope 3600 ft., one hen laid first egg on 13 May, another on 7 June.
1940	9.7	9 May	Very early year but hens nested in waves. 1st egg of latest hen 15 June.	5 May	Very early year but great spread.
1952	10.6	after 22 May		1 May	Most hens had laid by c.15 May.
1961	8.1	22 May	Most hens laid 4th week in May. Latest hen laid 1st egg on 15 June.	2 May	Most of nesting ground snow-free in first half of May. Appreciable plant growth in early May. (A. Watson)
1963	7.1	26 May	Average laying season.	5 May	Most of nesting ground snow-free in first half of May. Appreciable plant growth in early May. (A. Watson)

continued

TABLE 7   *continued*

Late Years

Year	Mean temperature (°C) for May at Braemar (1113 ft)	Dotterel		Ptarmigan	
		1st egg	Comments	1st egg	Comments
1934	8.3	1 June	A very rough May. Spread shorter than usual. Last hen laid 1st egg on 10 June.	23 May	A late season.
1951	6.9	after 21 June	Exceptionally heavy snow cover on high Cairngorms.	27 May	Many habitats deserted end of June. Second coldest May 1951-64. Exceptional snow cover delayed breeding.
1955	6.8	29 May	Coldest May 1951-64. Unusually heavy snow cover after mid-May.	27 May	Unusually heavy snow cover after mid-May. No plant growth or ptarmigan egg-laying in first half of May. (A. Watson)
1962	7.4	3 June	Heavy snow cover on high Cairngorms.	24 May	Average laying season on Lochnagar. (A. Watson)

TABLE 8 Clutch-size/Number of Clutches

Location	of 1	of 2	of 3	of 4	All	Average clutch size
Scotland, Cairngorms	1	16	178	0	195	2.91
East Grampians	0	0	29	0	29	3.0
Central and West Grampians	1	16	158	1	176	2.90
Other locations in Scotland	0	1	11	0	12	2.90
Scotland (all records)	2	33	376	1	412	2.91
England, Lakeland	4	16	38	0	58	2.58
Britain (all records)	6	49	414	1	470	2.87
Fenno-Scandia, Norway	0	5	81	1	87	2.95
Sweden	0	0	19	0	19	3.0
Finland	0	1	57	3	61	3.03
Fenno-Scandia (all records)	0	6	157	4	167	2.98
Continent of Europe Holland	0	0	12	0	12	(3.0)
Austria	0	0	7	0	7	(3.0)
Poland and Czechoslovakia	0	0	1	0	1	(3.0)

continued

TABLE 8 continued

Location	of 1	of 2	of 3	of 4	All	Average clutch size
Romania South and East Carpathians	0	0	2	0	2	(3.0)
Continent of Europe (all records)	0	0	20	0	20	3.0

TABLE 9 Percentage of nests with different clutch-sizes

Location	No. of nests	1 egg	2 eggs	3 eggs	4 eggs
Scotland	402	0.5	8.2	91.0	0.2
England	55	7.2	29.2	63.6	0
Great Britain	457	1.3	10.7	87.7	0.2
Fenno-Scandia	151	0	3.9	93.3	2.7

TABLE 10   CLUTCH SIZE   ENGLAND   (1784-1969)

Fells	1784-1899	1900-1926	1927-1969	Total	Mean
Buttermere Fells	1/3 1/2	2/3 6/2	1/2	11 3/3 8/2	2/2
Derwent Fells	-	1/1	-	1 (1/1)	(1.0)
Grassmoor	3/3	1/3 1/1	2/3	7 (6/3 1/1)	2.7
Skiddaw Range	8/3 1/2	1/2	1/3	11 9/3 2/2	2.8
Helvellyn Range	1/2	3/3 1/3 1/1	1/3	7 4/3 2/2 1/1	2.4
Cumberland (other fells)	3/3	4/3	5/3	12 12/3	3.0
Northumberland	-	1/3	-	1 (1/3)	(3.0)
Westmorland	1/3	3/3 1/2	-	5 4/3 1/2	2.8
Total	16/3 3/2	14/3 9/2 3/1	9/3 1/2	39/3 13/2 3/1	2.65
Mean	2.84 (19)	2.42 (26)	2.9 (10)	-	-

This Table gives a breakdown of all clutches recorded in England between 1784 - 1969.

D.

R

TABLE 11   CLUTCH SIZE   ENGLAND   (1900-1926)

Fell	1902	1905	1908	1911	1912	1914	1918	1920	1921	1922	1925	1926
Robinson			1/3 1/2	3/2	1/3		1/2	1/2				
Derwent Fells	1/1											
Grassmoor		1/1									1/3	
Skiddaw Range				1/2								
Helvellyn Range				1/3			1/1	1/2		1/3	1/3	
Cumberland						1/3			1/3		1/3	1/3
Westmorland									1/3		2/3	1/2
Northumberland				1/3								
Annual Total	1	1	2	6	1	1	2	2	2	1	5	2

Additional Notes:

    1921   2 or 3 pairs on Robinson.
    1922   1 bird on Robinson, more seen on Grassmoor.
    1923   None on Robinson, Grassmoor and Helvellyn.
    1924   None on Robinson, Grassmoor or Great Dun Fell.
    1928   None on Robinson or Helvellyn.

This Table gives a breakdown of clutch-size for the critical period between 1900-1926.

TABLE 12 REPLACEMENT CLUTCHES

No. of eggs in first clutch	Clutch taken or destroyed	Stage of incubation of first clutch when removed	No. of eggs in replacement clutch	Last egg laid in replacement clutch	Total period of replacement
3	14 June	fresh	3	22 June	8 days
3	8 June	6 days	2	17 June	9 days
3	9 June	c.7 days	2	20 June	11 days
3	12 June	fresh	3	21 June	9 days
3	4 June	12 days	3	2 July	19 days

TABLE 13 Egg laying dates of Ptarmigan and Dotterel in the Cairngorms and mean temperature in May at Braemar. Dates are given in days from 1st May.

Year	1st egg in 1st clutch Ptarmigan Cairngorms	1st egg in 1st clutch Dotterel Cairngorms (a)	1st egg in last clutch Dotterel Cairngorms (b)	Spread in Dotterel Cairngorms (b)-(a)	Mean temp. in May at Braemar (°F)
1933	6	12	34	22	47.6
1934	23	32	41	9	46.9
1935	18	37	no data	no data	45.1
1936	13	24	42	18	47.1
1937	11	20	27	7	48.3
1938	21	36	43	7	44.7
1939	18	31	36	5	46.9
1940	5	9	46	37	49.4
1941	13	27	38	11	44.5
1942	21	28	34	6	45.5
1945	15	29	37	8	46.8
1948	21	39	no data	no data	46.9
1949	18	21	49	28	47.1

continued

TABLE 13 *continued*

Year	1st egg in 1st clutch *Ptarmigan Cairngorms*	1st egg in 1st clutch *Dotterel Cairnforms (a)*	1st egg in last clutch *Dotterel Cairngorms (b)*	Spread in Dotterel Cairngorms (b)-(a)	Mean temp. in May at Braemar (°F)
1950	19	27	37	10	46.7
1951	27	53	62	9	44.4
1952	1	23	no data	no data	51.1
1955	27	29	52	23	44.2
1957	14	27	49	22	45.0
1958	20	31	42	11	44.7
1959	15	28	no data	no data	50.1
1960	21	22	30	8	49.3
1961	2	22	39	17	46.8
1962	24	34	42	8	45.3
1963	5	26	no data	no data	44.8
1970	no data	13	no data	no data	49.7

TABLE 14 Egg-laying dates in first clutches of Dotterel in the Grampians and English Lakeland

Year	Month	1st egg in 1st clutch		
		Grampians	Month	Lakeland
1908	May	16	May	19
1910	"	21	"	23
1911	"	14	"	11
1912	"	21	"	26
1914	"	23	"	29
1915	"	28	June	31
1925	"	19	May	27
1926	"	27	"	26
1959	"	23	"	28
1960	"	22	"	25

TABLE 15 Reproductive success of Dotterel in the Central Cairngorms in relation to the mean temperature in July at Braemar and to egg-laying dates of Dotterel. Dates given in days from 1st May.

Year	Young:old ratio	July mean temp. °F	1st egg date 1st clutch	1st egg date last clutch
1941	0.3	56.7	27	38
1942	0.3	54.3	28	34
1945	0.4	57.5	29	37
1947	0.3	56.9	28	no data
1962	0.2	51.9	34	42
1967	0.3*	55.2	no data	no data
1968	0.2*	54.0	no data	no data
1969	0.4*	56.8	no data	no data
1970	0.2*	52.3	no data	no data
1971	0.4*	55.7	no data	no data

* A. Watson's data

TABLE 16   INCUBATION PERIODS

No.	Location	Clutch completed	Hatch	Approx. period	Observer
1	Cairngorms	12 June. 3rd egg laid between 1209 and 1333.	7 July: 1813 2 chicks dry, 1 wet.	25.2	D.N.-T.
2	Cairngorms	10 June. 3rd egg laid c.1930.	8 July: 3 chicks in nest 2037.	27.5	D.N.-T.
3	Cairngorms	25 June. 3rd egg c.0700.	19 July: 3rd chick hatched between 1813-2021.	24.5	D.N.-T.
4	Cairngorms	6 June. 2nd egg laid between 1631-1942.	2 July: 1st chick hatched between 1717-2108.	26.1	D.N.-T.
5	Cairngorms	2 June. 3rd egg c.1809.	28 June: 3 chicks in nest 1421.	25.7	D.N.-T.
6	Cairngorms	3 June. 2nd egg laid.	2 July: 2 chicks, 1 wet.	29.7	Seton Gordon
7	Cairngorms	15 June. ♂ brooding 1 egg.	11 July: 4 hatched eggshells.	c.23	A. Tewnion
8	Zirbitzkögel, Austria	13 July. 3rd egg c.1745.	7 August: 3rd egg hatched 1150.	24.7	H. Franke

continued

TABLE 16   *continued*

No.	Location	Clutch completed	Hatch	Approx. period	Observer
9	Zirbitzkögel, Austria	3 June. 3rd egg between 0515-1630.	26 June: 2 eggs hatched within 4 hours. 1 egg addled.	24.2	H. Franke
10	Nordoost polder, Holland	8 May. 2nd egg: 10 May. 3rd egg.	3 June: 2 eggs hatched, 1 egg infertile.	c.26	J.F. Sollie
11	East Flevoland, Holland	26 May. 2 eggs.	23 July: 3 chicks.	c.27	M. Harpe
12-14	Finland	-		26-27 (3)	O. Hildén
15	Värriötunturi, Finland	-	-	27-28*	E. Pulliainen
16	Värriötunturi, Finland	-	-	25-26*	E. Pulliainen
17	Lakeland, England	c/3 17 May. 1st egg.c.14 May.	14 June: 1 egg and 1 new chick.	28-29	R. Stokoe

Average 26.1 days (17 clutches)

* same ♂ in successive years.

TABLE 17   CHICK MORTALITY IN SCOTLAND

Adults	Nests	Eggs laid	Eggs hatched	Young:old ratio at hatching	Young:old ratio end of 7 days
42	21	61	56	1.3:1	0.8:1

In 21 nests in the Cairngorms and Central Grampians 56 chicks hatched out of 61 eggs (90.8%), but only 33 (59%) survived until the end of the first week when there were 4 broods of 3 chicks, 6 of 2 chicks, 9 of 1 chick, and 2 with no chicks. The 5 addled eggs remained in the nest scrapes: 3 were infertile.

TABLE 18 BOREAL AND ARCTIC WADERS

Chick Patterns

Species	Hatching spread	Chick in nest	Parents	Flying period	Parents	Authority
Dotterel	usually 10-16 hours	4-16, up to 24 hours for older chick	♂ only normally broods and tends chicks.	26-30 days. Chicks can flutter about one week earlier.	♂ normally. ♀ later joins family group.	D.N.-T.
American Golden Plover	up to 34 hours	24 or more hours. Some leave nest, then return to be brooded.	Both sexes.	22 days	Both sexes.	D.F. Parmelee
Grey Plover	Often 24 to 48 hours	Up to 28 hours or longer, occasionally less than 16 hours.	Both sexes.	?	At first both sexes.	D.F. Parmelee
Ruddy Turnstone	11 to 15.8 hours	Leaves within 12 hours.	♂ stays with youngest nestling. ♀ moved off with 3 oldest.	c.20 days	At first both sexes. ♂ towards the end.	D.F. Parmelee
Baird's Sandpiper	?	Less than 24 hours	Both parents. ?♀ more solicitous.	c.19 days	At first both sexes. Parents desert chick before it flies.	D.F. Parmelee

TABLE 18 *continued*

Species	Hatching spread	Chick in nest	Parents	Flying period	Parents	Authority
Stilt Sandpiper	Within 20 hours	Over 12 hours	Both sexes at first. Later ♂ apparently takes over.	?	?	D.F. Parmelee
Grey Phalarope		Less than 24 hours	♂ only	c.18 days	♂ throughout fledging, but abandons even small chicks for hours at a time.	D.F. Parmelee

TABLE 19   NUMBERS (1900-71)

Region	Higher Numbers	Lower Numbers
	1900-1919	
ENGLAND	1908 (3), 1911 (7), 1912 (4), 1914 (3 plus), 1917 (4 plus)	None recorded 1900-01, 1904, 1907
SCOTLAND		
Grampians, East	1908, 1911, 1912	?
Central and West	1908, 1910, ?1915	1911
Cairngorms	?1914	?
North of Great Glen	1900-04, ?1912	?
	1920-29	
ENGLAND	1921 (8-9), 1925 (7), ?1926 (1-3)	1920, 1924, no pairs recorded 1927 1928, 1929
SCOTLAND		
Grampians, East	?	?
Central	1922, 1924, ?1927, 1929	1926
West	?	?
Cairngorms	1922, 1924, 1927, 1929	1926
North of Great Glen	1923, 1927	?
	1930-39	
ENGLAND	1932 (1), 1933 (1), 1934 (1), 1937 (1)	No nests recorded in other years
SCOTLAND		
Grampians, East	?	?
Central	1931, 1932, 1933, 1938	?1934, 1935
West	?	?
Cairngorms	1933, 1934, 1936, 1939	1931, 1932, 1935
North of Great Glen	?1935	?

continued

TABLE 19 *continued*

Region	*Higher Numbers*	*Lower Numbers*
	1940-49	
ENGLAND	?1944, 1945 (1)	No nests recorded 1946-49
SCOTLAND		
Grampians, East	?	?
Central	?1940	1942, 1945, 1946, 1949
West	?	?
Cairngorms	1940, ?1941, 1949	1942, 1948
North of Great Glen	?1941, 1948	
	1950-59	
ENGLAND	?1954, 1956 (1), 1959 (1)	No dotterels recorded 1951-53, 1955-58
SCOTLAND		
Grampians, East	?1951	?
Central	1951, 1953, 1954, 1958	?1952
West	1953, 1954	
Cairngorms	?1951, 1953, 1954	1950, 1956, 1957
North of Great Glen	1954, 1956, 1959	?
	1960-71	
ENGLAND	1960 (1), 1968 (1), 1969 (2),, 1970 (2), and 2 other birds, 1971 (1)	3 dotterels seen 1964, 1 1967. Other years no birds recorded
SCOTLAND		
Grampians, East	1963, 1966, 1969, 1970	(No records 1960-62), 1964, 1967
Central	1960, 1961	?1967 (E. of Drumochter)

continued

TABLE 19 *continued*

Region	Higher Numbers	Lower Numbers
	1960-71	
Grampians, West	1968	
Cairngorms	1960, 1961, 1967, 1969, 1970, 1971	1962, 1963, 1966, 1968
North of Great Glen	1967, ?1968, ?1969, 1970, ?1971	?

TABLE. 20　POPULATIONS AND BREEDING SUCCESS

Location	Year	Young reared	Adults ♂	Adults ♀	Total adults	Young/old ratio	Notes	Authority
England, Lakeland	1915	1	1	1	2	0.5 : 1	1 hill only. Apparently only 1 pair with fully-fledged juvenile.	G. Bolam
''	1916	2	1	1	2	1.0 : 1	Same hill. Young well fledged 8 Aug. Both adults accompanying young birds.	G. Bolam
Scotland, Cairngorms Central	1934	3	9	7	16	0.2 : 1	Egg collecting in addition to natural predation and hazards.	D.N.-T.
''	1936	2	7	6	13	0.2 : 1	''	D.N.-T.
''	1937	2			9	0.2 : 1	Ground incompletely covered. Some egg collecting.	D.N.-T.
''	1938	6			10	0.6 : 1	Ground not fully worked.	D.N.-T.

continued

TABLE 20 continued

Location	Year	Young reared	Adults ♂	Adults ♀	Total Adults	Young/old ratio	Notes	Authority
Scotland, Cairngorms Central	1940	5	7	7	14	0.4 : 1	Exceptionally early laying and hatching: warm June.	D.N.-T.
"	1941	3	5	7	12	0.3 : 1	Two surplus hens.	D.N.-T.
"	1949	4			21	0.2 : 1	Exceptionally heavy breeding stock.	D.N.-T.
"	1950	2			9	0.2 : 1	Only half 1949 breeding numbers.	D.N.-T.
"	1961	4			18-22	0.2 : 1	Year of high numbers. Ptarmigan nested exceptionally early.	B.N.-T.
"	1962	2	5	4	9	0.2 : 1	Smaller population than in 1961.	B.N.-T.
"	1967	5	8	8	16	0.3 : 1	Ground thoroughly covered.	A. Watson
"	1968	2	4	6	10	0.2 : 1	Very late thaw.	A. Watson

continued

S

TABLE 20 *continued*

Location	Year	Young reared	Adults ♂ ♀		Total adults	Young/old ratio	Notes	Authority
Cairngorms Central	1969	5	6	6	12	0.4 : 1	One ♀ lame. Ground thoroughly searched.	A. Watson
"	1970	5			29	0.2 : 1	Cold July.	A. Watson
"	1971	12			31	0.4 : 1	Warm July.	A. Watson
Cairngorms West	1940	2	3	2	5	0.4 : 1	Same ratio as Central Cairngorms in 1940.	D.N.-T.
"	1941	2	3	4	7	0.3 : 1	An extra ♀ sat on two infertile eggs.	D.N.-T.
"	1942	2	3	3	6	0.3 : 1		D.N.-T.
"	1945	3	3	?	7	0.4 : 1		D.N.-T.
"	1947	2			8	0.3 : 1		R. Perry

continued

TABLE 20 *continued*

Location	Year	Young reared	Adults ♂	Adults ♀	Total adults	Young/old ratio	Notes	Authority
Cairngorms East	1967	(A) 5	7	6	13	0.4 : 1	Two different hill groups.	A. Watson
"	1967	(B) 3	4	4	8	0.4 : 1		A. Watson
"	1968	(A) 3	7	9	16	0.2 : 1	Late thaw.	A. Watson
"	1969	(A) 3	2	2	4		Incomplete sample.	A. Watson
"	1970	(A) 3			19	0.2 : 1	Cold July.	A. Watson
Grampians East	1965	9			7	1.3 : 1	Ground thoroughly searched with pointers.	A. Watson
"	1966	9			9	1.0 : 1	"	A. Watson
"	1967	10			9	1.1 : 1	"	A. Watson M.J.P. Gregory

continued

TABLE 20 *continued*

Location	Year	Young reared	Adults ♂ ♀	Total adults	Young/old ratio	Notes	Authority
Grampians East	1968	5	3 4	7	0.7 : 1	Ground thoroughly searched with pointers.	A. Watson
"	1969	8	3 4	7	1.1 : 1	"	A. Watson
"	1970	10		10	1.0 : 1		A. Watson
Grampians Central	1953	14	17 14	31	0.5 : 1	Year of exceptionally high numbers. No egg collecting recorded.	D.N.-T.
"	1954	7		23	0.3 : 1		D.N.-T.
Grampians West	1962	1	1 1	2	0.5 : 1	Young not yet flying on 6 July.	A. Watson, Sr.
Sweden, Abisko	1958	3		6	0.5 : 1	Party of 6 adults including colour-banded pair with 3 young.	H. Rittinghaus 1962.

BREEDING SUCCESS

Summary				
Region	No. of years	Young reared	Adults	Young/old ratio
Northern England	2	3	4	0.8 : 1
Scotland, Grampians East	6	51	50	1.0 : 1
Grampians Central and West	3	22	56	0.4 : 1
Cairngorms East	5	17	60	0.3 : 1
Central	15	63	233	0.2 : 1
West	4	9	28	0.3 : 1
Cairngorms (All regions)		89	321	0.3 : 1
Sweden, Abisko	1	3	6	0.5 : 1

TABLE 21   EAST GRAMPIANS

Hill group	Year	Young reared	Adults ♂	Adults ♀	Total (adults)	Young/old ratio	Authority
A	1965	9			7	1.3 : 1	A. Watson
A	1966	9			9	1.0 : 1	A. Watson
A	1967	10	5	5	10	1.0 : 1	A. Watson and M.J.P. Gregory
A	1968	5	3	4	7	0.7 : 1	A. Watson
A	1969	8	3	4	7	1.1 : 1	A. Watson
A	1970	10			10	1.0 : 1	A. Watson
1965-70		51			50	1.0 : 1	as above

TABLE 22   Ptarmigan Numbers in Years of High Dotterel Populations

Central Grampians			
Year	Ptarmigan numbers	Authority	Comments
1898	?	J.D. MacKenzie	Small or doubtful peak in 1897
1908	Peak	"	
1910	Peak	"	
1915	?	"	
1922	Peak	"	
1927	Small peak	"	
1929	?	"	
1931	Peak	"	
1938	Peak	"	
1951	High	D.N.-T.	
1953	High	"	
1954	High	"	

TABLE 22

Cairngorms

1933	Very high	D.N.-T.
1934	Very high	"
1936	Dropping	"
1939	Not high	"
1940	High	"
1949	Rising	A. Watson
1951	Peak	"
1953	High	"
1954	Slump	"
1960	Rising	"
1961	High	"

(a) In the Central Grampians between 1898-1954 high breeding numbers of ptarmigan and dotterel coincided in at least 8 out of the 12 recorded good dotterel years.

(b) In the Cairngorms, between 1933-61, ptarmigan nested in high or rising numbers in only 7 of the 11 recorded good dotterel years.

(c) Ptarmigan, however, nested in high or rising numbers in the peak 'dotterel years' of 1910, 1922 and 1953 in the Grampians, and in 1933, 1934, 1940, 1949 and 1961 in the Cairngorms.

TABLE 23

Location	Sex	Habitat	Season	Food	Authority
England Lakeland	♂	nesting fell	summer	small coleopterous beetles	T.C. Heysham (1834)
"	♂	"	"	small *Diptera*. Craneflies *Nephrotoma*	E. Blezard
"	chick c.6 days	"	29.6.68	2 adult craneflies *Nephrotoma*. 4 carabid beetles *Calathus*. 3 weevils *Otiorrhynchus*. Finely ground fragments of *Coleoptera*. 3 angular fragments of quartz	E. Blezard
England Lincolnshire	?	farmlands	5 May ?	remains of *Coleoptera*. Larvae of *Lepidoptera* (Polyodon). Small particle of grit	J. Harting
Lancashire	7 birds	farmlands	early May	stomachs full of wireworms	F. Nicholson
Burgh-by-Sands, Cumberland	?	estuarine saltmarshes	15.5.1891	14 beetles *Chrysomela staphylea*. Fine grass shreds. Weevils *Sitona*. 23 particles of grit, mostly which quartz.	E. Blezard
"	3 birds	"	15.5.1891	stomachs of all 3 birds crammed with small beetles *Chrysomela hyperici*	H.A. MacPherson
Wigton, Cumberland	juvenile	under tele-graph wires	13.9.1932	2 carabid beetles *Pterostichus*. Nutlets of *Polygonum persicaria*	E. Blezard
Holland North-east polder	both	beetroot and potato fields	summer	insects and larvae. Never vegetable food	J.F. Sollie
Norway Dovre Fjell	?	nesting fells	June 1871	*Coleoptera*, chiefly *Bembidion* Weevils? *Otiorrhynchus* sp.	R. Collett
	?		May 1874	insects and larvae. Small leaves of *Salix*. Pieces of straw	R. Collett

continued

TABLE 23 *continued*

Location	Sex	Habitat	Season	Food	Authority
Sweden Abisko	both	nesting fells	1958-1961	insects in root systems of alpine grasslands. Apparently *not* on *Chironomidae*.	H. Rittinghaus
Finland Värriötunturi	both	nesting fells	summer	*Coleoptera, Diptera* and *Lepidoptera*. Blossoms of *Trientalis europaea, V. myrtillus*. Berries of *Empetrum nigrum*.	E. Pulliainen.
U.S.S.R.	both	nesting fells	summer	*Coleoptera* and larvae. *Diptera*. Occasionally seeds and berries of *Empetrum nigrum*.	A.I. Ivanov
"	"	on migration	spring and autumn	sometimes worms and molluscs.	A.I. Ivanov
Novaya Zemlya	"	tundra	summer	insects and large grains of quartz. Beetles.	S. Uspenskii
Lower Don	"	steppes	autumn	large quantities of larvae of beetles *Othous niger* and *Aniseplia austriaca* and caterpillars of butterfly *Cledobia moldavica*.	Pachkoskii
Scotland Central Grampians	both	nesting hills	summer	small *Diptera* and *Coleoptera* and larvae. Spiders.	D.N.-T.
Cairngorms	"	nesting hills	summer	*Tipulidae* and larvae. Hens feed on *tipulid* larvae in wet and mossy places and beside snow-fields, while cocks are brooding. *Diptera Coleoptera*. Spiders and harvestmen.	D.N.-T.
Southern Uplands	♂ and chick	nesting hill (stony summit)	June 1967	small *Diptera* and small *Coleoptera*. (Minute. Could not see what they took)	D. Watson

TABLE 24 NUMBERS IN ENGLAND (1900-28)

Year	Buttermere Fells	Derwent Fells	Skiddaw Group	Helvellyn Group	Cheviots	Cumberland	Westmorland	Yorkshire	Total for year
1900	-	-	-	-	-	-	-	-	-
1901	-	-	-	-	-	-	-	-	-
1902	-	1	-	-	-	-	-	-	1
1903	-	-	-	-	-	-	-	-	-
1904	-	-	-	-	-	-	-	-	-
1905	-	1	-	-	-	-	-	1	2
1906	1	-	-	-	-	-	-	-	1
1907	-	-	-	-	-	-	-	-	-
1908	3	-	-	-	-	-	-	-	3
1909	x^2	-	-	-	-	-	-	-	x^2
1910	2+	-	-	-	-	-	-	-	2+
1911	4	-	1	1	1	-	-	-	7
1912	1	-	-	?2	1	-	-	-	?4.
1913	x^2	-	-	-	-	1	-	-	1&x^2
1914	-	-	-	x	-	3+	-	-	3+
1915	-	-	-	-	-	1	-	-	1
1916	-	-	-	-	-	1	-	-	1
1917	0	-	0	2+	-	2+	-	-	4+
1918	0	-	0	2	-	-	-	-	2
1919	1	-	-	-	-	-	-	-	1
1920	1	-	-	-	-	-	-	-	1
1921	2-3	-	-	-	-	5+	-	-	7+
1922	x0	-	-	1	-	1	-	-	2
1923	x^2	-	0	0	-	2	-	-	2&x^2
1924	02	-	-	x	-	-	-	-	x&02
1925	2	-	-	-	-	?5	-	-	?7
1926	-	-	-	-	-	1-3	-	-	1-3
1927	-	-	-	x	-	0	-	-	x0
1928	0	-	x	0	-	-	-	-	x0

Notes

Numbers = pairs recorded in summer on nesting ground

x = single bird recorded in summer on nesting ground

0 = record of fell searched but no dotterel recorded

- = no record

Index

I have not indexed the names of plants or their scientific names. Botanists should read Derek Ratcliffe's detailed chapter on the dotterel's breeding habitat in Britain (pp. 153-73).